USA and International Mathematical Olympiads 2000

© 2001 by
The Mathematical Association of America (Incorporated)
Library of Congress Catalog Card Number 2001090644
ISBN 0-88385-804-5
Printed in the United States of America
Current Printing (last digit):
10 9 8 7 6 5 4 3 2 1

USA and International Mathematical Olympiads 2000

Edited by
Titu Andreescu
and
Zuming Feng

Published and distributed by
The Mathematical Association of America

MAA PROBLEM BOOKS SERIES

Problem Books is a series of the Mathematical Association of America consisting of collections of problems and solutions from annual mathematical competitions; compilations of problems (including unsolved problems) specific to particular branches of mathematics; books on the art and practice of problem solving, etc.

Committee on Publications
William Watkins, *Chair*

Roger Nelsen *Editor*
Irl Bivens   Clayton Dodge
Richard Gibbs   George Gilbert
Art Grainger   Loren Larson
Margaret Robinson

Mathematical Olympiads 1998–1999: Problems and Solutions From Around the World, edited by Titu Andreescu and Zuming Feng

USA and International Mathematical Olympiads 2000, edited by Titu Andreescu and Zuming Feng

MAA Service Center
P. O. Box 91112
Washington, DC 20090-1112
1-800-331-1622   fax: 1-301-206-9789

Preface

This book is intended to help students preparing to participate in the USA Mathematical Olympiad (USAMO) in the hope of representing the United States at the International Mathematical Olympiad (IMO). The USAMO is the third stage of the selection process leading to participation in the IMO. The preceding examinations are the AMC 10 or AMC 12 (which replaced the American High School Mathematics Examination) and the American Invitational Mathematics Examination (AIME). Participation in the AIME and the USAMO is by invitation only, based on performance in the preceding exams of the sequence.

Newly introduced in 2000 was the Team Selection Test. The top 12 USAMO students are invited to attend the Mathematical Olympiad Summer Program (MOSP) regardless of their grade. An additional 12–18 students MOSP invitations are extended to the most promising non-graduating USAMO students, potential IMO participants in future years. During the first week of MOSP a final IMO type exam is given to the top 12 USAMO students with the goal of identifying the USA IMO Team. This exam replicates an actual IMO, consisting of 6 problems to be solved over two 4 1/2 hour sessions. The 12 equally weighted problems (6 on the USAMO and 6 on this exam) determine the tentative USA Team.

The Mathematical Olympiad booklets have been published since 1976. Copies are $5.00 (payable in U.S. dollars) for each year from 1976 through 1999. This is the first volume offered in book form. In addition, various other publications are useful in preparing for the AMC-AIME-USAMO-IMO sequence (see the reading list that starts on page 71). For further information about American Mathematics Competitions' examinations, or to order Mathematical Olympiad booklets from additional years, please write to

Titu Andreescu, Director
American Mathematics Competitions
University of Nebraska-Lincoln
1740 Vine St.
Lincoln, NE 68588-0658,

or visit the AMC web site at www.unl.edu/amc.

Contents

Preface   v

Acknowledgments   ix

Abbreviations and Notations   xi

Introduction   xiii

1 The Problems   **1**
    1   USAMO 1
    2   Team Selection Test 2
    3   IMO 4

2 Hints   **7**
    1   USAMO 7
    2   Team Selection Test 8
    3   IMO 8

3 Formal Solutions   **10**
    1   USAMO 10
    2   Team Selection Test 23
    3   IMO 33

4 Problem Credits   **63**

5 Glossary   **65**

6 Further Reading   **71**

Appendix **75**
1. 2000 Olympiad Results 75
2. 1999 Olympiad Results 77
3. 1998 Olympiad Results 78
4. 1997 Olympiad Results 79
5. 1996 Olympiad Results 80
6. 1996–2000 Cumulative IMO Results 82

Acknowledgments

Thanks to Kiran Kedlaya and George Lee who helped in ways too numerous to mention. Without their efforts this work would not have been possible. Thanks also to Reid Barton and Ricky Liu, who helped in preparing solutions to the IMO problems 4 and 5.

Abbreviations and Notations

Abbreviations

IMO	International Mathematical Olympiad
USAMO	United States of America Mathematical Olympiad
MOSP	Mathematical Olympiad Summer Program

Notations for Numerical Sets and Fields

$\mathbb{Z}$	the set of integers
$\mathbb{Z}^+$	the set of positive integers
$\mathbb{Z}_n$	the set of integers modulo n
$\mathbb{Q}$	the set of rational numbers
$\mathbb{Q}^+$	the set of positive rational numbers
$\mathbb{Q}^n$	the set of n-tuples of rational numbers
$\mathbb{R}$	the set of real numbers
$\mathbb{R}^+$	the set of positive real numbers
$\mathbb{R}^n$	the set of n-tuples of real numbers
$\mathbb{C}$	the set of complex numbers

Notations for Sets, Logic, and Geometry

$\iff$	if and only if
$\implies$	implies
$\|A\|$	the number of elements in set A
$A \subset B$	A is a proper subset of B
$A \subseteq B$	A is a subset of B
$A \setminus B$	A without B
$A \cap B$	the intersection of sets A and B
$A \cup B$	the union of sets A and B
$a \in A$	the element a belongs to the set A
$\overline{AB}$	segment AB
AB	the length of $\overline{AB}$
$\overparen{AB}$	arc AB
$\overrightarrow{AB}$	vector AB
$[\mathcal{F}]$	area of figure $\mathcal{F}$

Introduction

Olympiad-style exams consist of several challenging essay problems. Correct solutions often require deep analysis and careful argument. Olympiad questions can seem impenetrable to the novice, yet most can be solved with elementary high school mathematics, cleverly applied.

Here is some advice for students who attempt the problems that follow.

- Take your time! Very few contestants can solve all of the given problems within the time limit. Ignore the time limit if you wish.
- Try the "easier" questions first (problems 1 and 4 on each exam).
- Olympiad problems don't "crack" immediately. Be patient. Try different approaches. Experiment with simple cases. In some cases, working backward from the desired result is helpful.
- If you get stumped, glance at the *Hints* section. Sometimes a problem requires an unusual idea or exotic technique, which might be explained in this section.
- Even if you can solve a problem, read the hints and solutions. The hints may contain some ideas that did not occur in your solution, and they may discuss strategic and tactical approaches that can be used elsewhere. The formal solutions are models of elegant presentation that you should emulate, but they often obscure the torturous process of investigation, false starts, inspiration and attention to detail that led to them. When you read the formal solutions, try to reconstruct the thinking that went into them. Ask yourself, "What were the key ideas?" "How can I apply these ideas further?"
- Go back to the original problem later, and see if you can solve it in a different way. Many of the problems have multiple solutions, but not

all are outlined here.

- All terms in boldface are defined in the *Glossary*. Use the glossary and the reading list to further your mathematical education.
- Meaningful problem solving takes practice. Don't get discouraged if you have trouble at first. For additional practice, use prior years' exams or the books on the reading list.

1
The Problems

1 USAMO

29th United States of America Mathematical Olympiad
Part I   9 A.M.–12 P.M.
May 2, 2000

1. Call a real-valued function f *very convex* if
$$\frac{f(x)+f(y)}{2} \geq f\left(\frac{x+y}{2}\right) + |x-y|$$
holds for all real numbers x and y. Prove that no very convex function exists.

2. Let S be the set of all triangles ABC for which
$$5\left(\frac{1}{AP}+\frac{1}{BQ}+\frac{1}{CR}\right) - \frac{3}{\min\{AP,BQ,CR\}} = \frac{6}{r},$$
where r is the inradius and P,Q,R are the points of tangency of the incircle with sides AB, BC, CA, respectively. Prove that all triangles in S are isosceles and similar to one another.

3. A game of solitaire is played with R red cards, W white cards, and B blue cards. A player plays all the cards one at a time. With each play he accumulates a penalty. If he plays a blue card, then he is charged a penalty which is the number of white cards still in his hand. If he plays a white card, then he is charged a penalty which is twice the number of red cards still in his hand. If he plays a red card, then he is

charged a penalty which is three times the number of blue cards still in his hand. Find, as a function of R, W, and B, the minimal total penalty a player can amass and all the ways in which this minimum can be achieved.

29th United States of America Mathematical Olympiad

Part II   1 P.M.–4 P.M.

May 2, 2000

4. Find the smallest positive integer n such that if n unit squares of a 1000×1000 unit-square board are colored, then there will exist three colored unit squares whose centers form a right triangle with legs parallel to the edges of the board.

5. Let $A_1 A_2 A_3$ be a triangle and let ω_1 be a circle in its plane passing through A_1 and A_2. Suppose there exist circles $\omega_2, \omega_3, \ldots, \omega_7$ such that for $k = 2, 3, \ldots, 7$, ω_k is externally tangent to ω_{k-1} and passes through A_k and A_{k+1}, where $A_{n+3} = A_n$ for all $n \geq 1$. Prove that $\omega_7 = \omega_1$.

6. Let $a_1, b_1, a_2, b_2, \ldots, a_n, b_n$ be nonnegative real numbers. Prove that

$$\sum_{i,j=1}^{n} \min\{a_i a_j, b_i b_j\} \leq \sum_{i,j=1}^{n} \min\{a_i b_j, a_j b_i\}.$$

2 Team Selection Test

41st IMO Team Selection Test

Lincoln, Nebraska

Day I   11:30 A.M.–4 P.M.

June 10, 2000

1. Let a, b, c be nonnegative real numbers. Prove that

$$\frac{a+b+c}{3} - \sqrt[3]{abc} \leq \max\{(\sqrt{a} - \sqrt{b})^2, (\sqrt{b} - \sqrt{c})^2, (\sqrt{c} - \sqrt{a})^2\}.$$

2. Let $ABCD$ be a cyclic quadrilateral and let E and F be the feet of perpendiculars from the intersection of diagonals AC and BD to $\overline{AB}$ and $\overline{CD}$, respectively. Prove that $\overline{EF}$ is perpendicular to the line through the midpoints of $\overline{AD}$ and $\overline{BC}$.

3. For real number x, let $\lceil x \rceil$ denote the smallest integer greater than or equal to x, let $\lfloor x \rfloor$ denote the greatest integer less than or equal to x, and let $\{x\}$ denote the fractional part of x, which is given by $x - \lfloor x \rfloor$. Let p be a prime number. For integers r, s such that $rs(r^2 - s^2)$ is not divisible by p, let $f(r, s)$ denote the number of integers $n \in \{1, 2, \ldots, p-1\}$ such that $\{rn/p\}$ and $\{sn/p\}$ are either both less than $1/2$ or both greater than $1/2$. Prove that there exists $N > 0$ such that for $p \geq N$ and all r, s,

$$\left\lceil \frac{p-1}{3} \right\rceil \leq f(r, s) \leq \left\lfloor \frac{2(p-1)}{3} \right\rfloor.$$

41st IMO Team Selection Test

Lincoln, Nebraska

Day II   11:30 A.M.–4 P.M.

June 11, 2000

4. Let n be a positive integer. Prove that

$$\sum_{i=0}^{n} \binom{n}{i}^{-1} = \frac{n+1}{2^{n+1}} \sum_{i=1}^{n+1} \frac{2^i}{i}.$$

5. Let n be a positive integer. A *corner* is a finite set C of ordered n-tuples of positive integers such that if $a_1, a_2, \ldots, a_n, b_1, b_2, \ldots, b_n$ are positive integers with $a_k \geq b_k$ for $k = 1, 2, \ldots, n$ and $(a_1, a_2, \ldots, a_n) \in C$, then $(b_1, b_2, \ldots, b_n) \in C$. Prove that among any infinite collection S of corners, there exist two corners, one of which is a subset of the other one.

6. Let ABC be a triangle inscribed in a circle of radius R, and let P be a point in the interior of ABC. Prove that

$$\frac{PA}{BC^2} + \frac{PB}{CA^2} + \frac{PC}{AB^2} \geq \frac{1}{R}.$$

3 IMO

41st International Mathematical Olympiad
Taejon, Republic of Korea
Day I 9 A.M.–1:30 P.M.
July 19, 2000

1. Two circles ω_1 and ω_2 intersect at M and N. Line ℓ is tangent to the circles at A and B, respectively, so that M lies closer to ℓ than N. Line CD, with C on ω_1 and D on ω_2, is parallel to ℓ and passes through M. Let lines AC and BD meet at E; let lines AN and CD meet at P; and let lines BN and CD meet at Q. Prove that $EP = EQ$.

2. Let a, b, c be positive real numbers such that $abc = 1$. Prove that
$$(a - 1 + 1/b)(b - 1 + 1/c)(c - 1 + 1/a) \leq 1.$$

3. Let $n \geq 2$ be a positive integer. Initially, there are n fleas on a horizontal line, not all at the same point. For a positive real number λ, define a *move* as follows:

    choose any two fleas, at points A and B, with A to the left of B; let the flea at A jump to the point C on the line to the right of B with $BC/AB = \lambda$.

    Determine all values of λ such that, for any point M on the line and any initial positions of the n fleas, there is a finite sequence of moves that will take all the fleas to positions to the right of M.

41st International Mathematical Olympiad
Taejon, Republic of Korea
Day II 9 A.M.–1:30 P.M.
July 20, 2000

4. A magician has one hundred cards numbered 1 to 100. He puts them into three boxes, a red one, a white one and a blue one, so that each box contains at least one card. A member of the audience selects two

of the three boxes, chooses one card from each and announces the sum of the numbers on the chosen cards. Given this sum, the magician identifies the box from which no card has been chosen. How many ways are there to put all the cards into the boxes so that this trick always works? (Two ways are considered different if at least one card is put into a different box.)

5. Determine if there exists a positive integer n such that n has exactly 2000 prime divisors and $2^n + 1$ is divisible by n.

6. Let $\overline{AH_1}$, $\overline{BH_2}$, and $\overline{CH_3}$ be the altitudes of an acute triangle ABC. The incircle ω of triangle ABC touches the sides BC, CA, and AB at T_1, T_2, and T_3, respectively. Consider the symmetric images of the lines H_1H_2, H_2H_3, and H_3H_1 with respect to the lines T_1T_2, T_2T_3, and T_3T_1. Prove that these images form a triangle whose vertices lie on ω.

2
Hints

I USAMO

1. Try iterations and telescoping sums. Also see if you can deduce an inequality similar to the given one but with a different expression in place of $|x - y|$.

2. Prove a useful trigonometric identity:

$$\tan\frac{A}{2}\tan\frac{B}{2} + \tan\frac{B}{2}\tan\frac{C}{2} + \tan\frac{C}{2}\tan\frac{A}{2} = 1,$$

   where A, B, C are the angles of triangle ABC.

   You will discover that you have only two equations in three variables, and two of them are symmetric to each other. In order to determine the angles from these, you will have to take advantage of a suitable substitution and inequality.

3. Given a hand, how does one make changes step by step to reduce the total penalty?

4. First find an example with as many squares colored as possible having no such right triangle. Then prove that your example is best possible either by induction (but be careful with your base cases!) or by a counting argument.

5. Use common tangent lines to express the angles of the triangle in terms of the arcs on the circles.

6. Try varying one of the variables while keeping the others fixed and see if you can make the difference between the two sides smaller. If this is impossible, you may be able to simplify the problem by combining

some terms: If $a_i/b_i = a_j/b_j$, note how replacing (a_i, b_i) and (a_j, b_j) with (a_i+a_j, b_i+b_j) affects the inequality. Find a similar replacement when $a_i/b_i = b_j/a_j$. Also notice that if $a_i < b_i$, then within a certain interval the expressions are *linear* with respect to b_i.

2 Team Selection Test

1. Make an assumption without loss of generality, and/or replace the maximum by the arithmetic mean to simplify the right side.

2. Use other midpoints to build useful midlines.

3. First reduce to $r = 1$, and notice that $s = \pm 3$ give equality cases. Also note that it suffices to count up to $(p-1)/2$ and double the result to get $f(r, s)$. Now try to control the difference between the number of times elements of the sequence $\{s/p\}, \{2s/p\}, \ldots, \{(p-1)s/p\}$ lie in the intervals $(0, 1/2)$ and $(1/2, 1)$. It will help to consider "small" and "large" s separately.

4. Induction! Play with **binomial coefficients** by regrouping effectively.

5. It might help to first try the case $n = 2$, or to first try the case where each corner is a single rectangle. (But don't forget that an arbitrary corner need not be rectangular!) If you try induction on n, you may need to induct on the stronger statement that any infinite sequence of corners contains an increasing subsequence. In any case, writing a complete proof in general will test your ability to present sophisticated mathematical arguments at a professional level.

6. The question resembles the famous **Erdős-Mordell** inequality. The proof of this inequality involves relating the length of a segment to its projection onto another line, then judiciously combining several such relations.

3 IMO

1. Try to find a **kite**.

2. How can you use $abc = 1$ effectively? Try to perform symmetric substitutions or to get rid of the fractions in different ways.

Hints

3. First determine the critical value for λ by trying some simple cases, say $n = 2, 3, 4$. Then try to find some useful monovariants (functions that don't increase when a move is made) related to the positions of the fleas.

4. The sum of the cards from distinct boxes must be unique. By the way, the answer is not 6.

5. Induction! Find a positive integer n such that n has exactly i, $i = 1, 2, 3, \ldots$, prime divisors and $2^n + 1$ is divisible by n.

6. First prove that there is a **homothety** between the newly formed triangle and the original triangle. Determine the ratio and the center of the homothety.

3
Formal Solutions

I USAMO

1. Call a real-valued function f *very convex* if
$$\frac{f(x)+f(y)}{2} \geq f\left(\frac{x+y}{2}\right) + |x-y|$$
holds for all real numbers x and y. Prove that no very convex function exists.

First Solution. Fix $n \geq 1$. For each integer i, define
$$\Delta_i = f\left(\frac{i+1}{n}\right) - f\left(\frac{i}{n}\right).$$
The given inequality with $x = (i+2)/n$ and $y = i/n$ implies
$$\frac{f\left(\frac{i+2}{n}\right) + f\left(\frac{i}{n}\right)}{2} \geq f\left(\frac{i+1}{n}\right) + \frac{2}{n},$$
from which,
$$f\left(\frac{i+2}{n}\right) - f\left(\frac{i+1}{n}\right) \geq f\left(\frac{i+1}{n}\right) - f\left(\frac{i}{n}\right) + \frac{4}{n},$$
that is, $\Delta_{i+1} \geq \Delta_i + 4/n$. Combining this for n consecutive values of i gives $\Delta_{i+n} \geq \Delta_i + 4$. Summing this inequality for $i = 0$ to $i = n-1$ and cancelling terms yields
$$f(2) - f(1) \geq f(1) - f(0) + 4n.$$
This cannot hold for all $n \geq 1$. Hence there are no very convex functions.

Formal Solutions

Second Solution. We show by induction that the given inequality implies
$$\frac{f(x) + f(y)}{2} - f\left(\frac{x+y}{2}\right) \geq 2^n |x - y|$$
for all nonnegative integers n. This will yield a contradiction, because for fixed x and y the right side gets arbitrarily large, while the left side remains fixed.

We are given the base case $n = 0$. Now if the inequality holds for a given n, then for a, b real,
$$\frac{f(a) + f(a+2b)}{2} \geq f(a+b) + 2^{n+1}|b|,$$
$$f(a+b) + f(a+3b) \geq 2(f(a+2b) + 2^{n+1}|b|),$$
and
$$\frac{f(a+2b) + f(a+4b)}{2} \geq f(a+3b) + 2^{n+1}|b|.$$
Adding these three inequalities and canceling terms yields
$$\frac{f(a) + f(a+4b)}{2} \geq f(a+2b) + 2^{n+3}|b|.$$
Setting $x = a$, $y = a + 4b$, we obtain
$$\frac{f(x) + f(y)}{2} \geq f\left(\frac{x+y}{2}\right) + 2^{n+1}|x - y|,$$
and the induction is complete.

Third Solution. Rewrite the condition as
$$f(x) - f\left(\frac{x+y}{2}\right) \geq f\left(\frac{x+y}{2}\right) - f(y) + 2|x - y|.$$
For any positive integer n,
$$f(1) - f(0) = f(1) - f(1/2) + f(1/2) - f(0)$$
$$\geq f(1/2) - f(0) + 2 + f(1/2) - f(0) = 2[f(1/2) - f(0)] + 2$$
$$= 2[f(1/2) - f(1/4) + f(1/4) - f(0)] + 2$$
$$\geq 2[f(1/4) - f(0) + 1 + f(1/4) - f(0)] + 2$$
$$= 4[f(1/4) - f(0)] + 4$$
$$= \cdots \geq 2^n[f(1/2^n) - f(0)] + 2n.$$

Similarly, $f(-1) - f(0) \geq 2^n[f(-1/2^n) - f(0)] + 2n$. But

$$f(1/2^n) + f(-1/2^n) \geq 2f(0) + 1/2^{n-2} > 2f(0).$$

Thus, for each n, at least one of $f(1/2^n) - f(0)$ and $f(-1/2^n) - f(0)$ is greater than 0. It follows that at least one of $f(1) - f(0)$ and $f(-1) - f(0)$ is greater than $2n$ for all $n \geq 1$, which is impossible. Hence there is no very convex functions.

2. Let S be the set of all triangles ABC for which

$$5\left(\frac{1}{AP} + \frac{1}{BQ} + \frac{1}{CR}\right) - \frac{3}{\min\{AP, BQ, CR\}} = \frac{6}{r},$$

where r is the inradius and P, Q, R are the points of tangency of the incircle with sides AB, BC, CA, respectively. Prove that all triangles in S are isosceles and similar to one another.

First Solution. We start with the following lemma.

Lemma. Let A, B, C be the angles of triangle ABC. Then

$$\tan\frac{A}{2}\tan\frac{B}{2} + \tan\frac{B}{2}\tan\frac{C}{2} + \tan\frac{C}{2}\tan\frac{A}{2} = 1,$$

Proof. We present arguments

First approach. Since

$$\tan(\alpha + \beta)[1 - \tan\alpha\tan\beta] = \tan\alpha + \tan\beta,$$

$\tan(90° - \alpha) = \cot\alpha = 1/\tan\alpha$, and $A/2 + B/2 + C/2 = 90°$, the desired follows from

$$\tan\frac{A}{2}\tan\frac{B}{2} + \tan\frac{B}{2}\tan\frac{C}{2}$$

$$= \tan\frac{B}{2}\left(\tan\frac{A}{2} + \tan\frac{C}{2}\right)$$

$$= \tan\frac{B}{2}\tan\left(\frac{A}{2} + \frac{C}{2}\right)\left[1 - \tan\frac{A}{2}\tan\frac{C}{2}\right]$$

$$= \tan\frac{B}{2}\tan\left(90° - \frac{B}{2}\right)\left[1 - \tan\frac{A}{2}\tan\frac{C}{2}\right]$$

$$= 1 - \tan\frac{A}{2}\tan\frac{C}{2}.$$

Second approach. Let a, b, c, r, s denote the side lengths, inradius and semiperimeter of triangle ABC, respectively. Then $[ABC] = rs$, $AP = s - a$, and $\tan(A/2) = r/(s-a)$. Hence

$$\tan\left(\frac{A}{2}\right) = \frac{[ABC]}{s(s-a)}.$$

Likewise,

$$\tan\left(\frac{B}{2}\right) = \frac{[ABC]}{s(s-b)} \text{ and } \tan\left(\frac{C}{2}\right) = \frac{[ABC]}{s(s-c)}.$$

Hence

$$\tan\frac{A}{2}\tan\frac{B}{2} + \tan\frac{B}{2}\tan\frac{C}{2} + \tan\frac{C}{2}\tan\frac{A}{2}$$
$$= \frac{[ABC]^2}{s^2}\left(\frac{(s-c)+(s-a)+(s-b)}{(s-a)(s-b)(s-c)}\right)$$
$$= \frac{[ABC]^2}{s(s-a)(s-b)(s-c)} = 1,$$

by **Heron's formula**. □

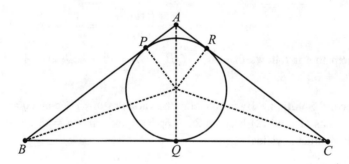

Without loss of generality assume that $AP = \min\{AP, BQ, CR\}$. Let $x = \tan(\angle A/2)$, $y = \tan(\angle B/2)$, and $z = \tan(\angle C/2)$. Then $AP = r/x$, $BQ = r/y$, and $CR = r/z$. Then the equation given in the problem statement becomes

$$2x + 5y + 5z = 6, \tag{1}$$

and the equation in the lemma is

$$xy + yz + zx = 1. \tag{2}$$

Eliminating x from (1) and (2) yields
$$5y^2 + 5z^2 + 8yz - 6y - 6z + 2 = 0.$$
Completing squares, we obtain
$$(3y-1)^2 + (3z-1)^2 = 4(y-z)^2.$$
Setting $3y - 1 = u$, $3z - 1 = v$ (i.e., $y = (u+1)/3$, $z = (v+1)/3$) gives
$$5u^2 + 8uv + 5v^2 = 0.$$
Because the discriminant of this quadratic equation is $8^2 - 4 \times 25 < 0$, the only real solution to the equation is $u = v = 0$. Thus there is only one possible set of values for the tangents of half-angles of ABC (namely $x = 4/3$, $y = z = 1/3$). Thus all triangles in S are isosceles and similar to one another.

Indeed, we have $x = r/AP = 4/3$ and $y = z = r/BQ = r/CQ = 1/3 = 4/12$, so we can set $r = 4$, $AP = AR = 3$, and $BP = BQ = CQ = CR = 12$. This leads to $AB = AC = 15$ and $BC = 24$. By scaling, all triangles in S are similar to the triangle with side lengths 5, 5, 8.

We can also use **half-angle formulas** to calculate
$$\sin B = \sin C = \frac{2\tan\frac{C}{2}}{1 + \tan^2\frac{C}{2}} = \frac{3}{5}.$$
From this it follows that $AQ : QB : BA = 3 : 4 : 5$ and $AB : AC : BC = 5 : 5 : 8$.

Second Solution. By introducing the variables $p = y + z$ and $q = yz - 1$, relations (1) and (2) become $2x + 5p = 6$ and $xp + q = 0$, respectively. Eliminating x yields
$$p(6 - 5p) + 2q = 0. \tag{3}$$
Note that y and z are the roots of the equation
$$t^2 - pt + (q + 1) = 0. \tag{4}$$
Expressing q in terms of p in (3), and substituting in (4), we obtain the following quadratic equation in t:
$$t^2 - pt + \frac{5p^2 - 6p + 2}{2} = 0.$$
This equation has discriminant $-(3p - 2)^2 \leq 0$. Hence the equation has real solutions only if $p = 2/3$, and $y = z = 1/3$.

Note. We can also let $x = AP$, $y = BQ$, $z = CR$ and use the fact that
$$r(x+y+z) = [ABC] = \sqrt{xyz(x+y+z)}$$
to obtain a quadratic equation in three variables. Without loss of generality, we may set $x = 1$. Then the solution proceeds as above.

3. A game of solitaire is played with R red cards, W white cards, and B blue cards. A player plays all the cards one at a time. With each play he accumulates a penalty. If he plays a blue card, then he is charged a penalty which is the number of white cards still in his hand. If he plays a white card, then he is charged a penalty which is twice the number of red cards still in his hand. If he plays a red card, then he is charged a penalty which is three times the number of blue cards still in his hand. Find, as a function of R, W, and B, the minimal total penalty a player can amass and all the ways in which this minimum can be achieved.

Solution. The minimum achievable penalty is
$$\min\{BW, 2WR, 3RB\}.$$
The three penalties BW, $2WR$, and $3RB$ can clearly be obtained by playing cards in one of the three orders

- bb $\cdots$ brr $\cdots$ rww $\cdots$ w,
- rr $\cdots$ rww $\cdots$ wbb $\cdots$ b,
- ww $\cdots$ wbb $\cdots$ brr $\cdots$ r.

Given an order of play, let a "run" of some color denote a set of cards of that color played consecutively in a row. Then the optimality of one of the three above orders follows immediately from the following lemma, along with the analogous observations for blue and white cards.

Lemma 1. For any given order of play, we may combine any two runs of red cards without increasing the penalty.

Proof. Suppose that there are w white cards and b blue cards between the two red runs. Moving a red card from the first run to the second costs us $2w$ because we now have one more red card after the w white cards. However, we gain $3b$ because this red card is now after the b blues. If the net gain $3b - 2w$ is non-negative, then we can move all the red cards in the first run to the second run without increasing the penalty. If the net gain $3b - 2w$ is negative, then we can move all the red cards in the second run to the first run without increasing the penalty, as desired. □

Thus there must be an optimal game where cards are played in one of the three given orders. To determine whether there are other optimal orders, first observe that **wr** can never appear during an optimal game; otherwise, playing these two cards in the order **rw** instead decreases the penalty. Similarly, **bw** and **rb** can never appear. Now we prove the following lemma.

Lemma 2. Any optimal order of play must have less than 5 runs.

Proof. Suppose that some optimal order of play had at least five runs. Assume the first card played is red; the proof is similar in the other cases. Say we first play r_1, w_1, b_1, r_2, w_2 cards of each color, where each r_i, w_i, b_i is positive and where we cycle through red, white, and blue runs. From the proof of our first lemma we must have both $3b_1 - 2w_1 = 0$ and $b_1 - 2r_2 = 0$. Hence the game starting with playing $r_1, w_1 + w_2, b_1, r_2$, 0 cards is optimal as well, so we must also have $3b_1 - 2(w_1 + w_2) = 0$, a contradiction. □

Thus, any optimal game has at most 4 runs. Now from lemma 1 and our initial observations, any order of play of the form

$$\mathbf{rr} \cdots \mathbf{rww} \cdots \mathbf{wbb} \cdots \mathbf{brr} \cdots \mathbf{r},$$

is optimal if and only if $2W = 3B$ and $2WR = 3RB \leq WB$; and similar conditions hold for 4-run games that start with **w** or **b**.

4. Find the smallest positive integer n such that if n unit squares of a 1000×1000 unit-square board are colored, then there will exist three colored unit squares whose centers form a right triangle with legs parallel to the edges of the board.

First Solution. We show that $n = 1999$. Indeed, $n \geq 1999$ because we can color 1998 squares without producing a right triangle: color every square in the first row and the first column, except for the one square at their intersection.

Now assume that some squares have been colored so that no desired right triangle is formed. Call a row or column *heavy* if it contains more than one colored square, and *light* otherwise. Our assumption then states that no heavy row and heavy column intersect in a colored square.

If there are no heavy rows, then each row contains at most one colored square, so there are at most 1000 colored squares. We reach the same conclusion, if there are no heavy columns. If there is a heavy row and a heavy column, then by the initial observation, each colored square in the heavy row or column must lie in a light column or row, and no two can lie in the same light column or row. Thus the number of colored squares is at most the number of light rows and columns, which is at most $2 \times (1000 - 1) = 1998$.

We conclude that in fact 1999 colored squares is the minimum needed to force the existence of a right triangle of the type described.

Second Solution. Assume that 1999 squares are colored and the required right triangle does not exist. By the Pigeonhole Principle, there is a row with $a_1 \geq 2$ colored squares. Interchange rows to make this the first row. Interchange columns so that the first a_1 squares in the first row are all colored. Then the first a_1 columns have no colored squares other than the ones in the first row, for otherwise we would have a right triangle.

Observe that a_1 cannot equal 1000, for then we would have no place for the remaining 999 colored squares. Also, a_1 cannot equal 999, for then the remaining 1000 colored squares must all be in the last column and we would have a right triangle, a contradiction. Hence $1000 - a_1 \geq 2$.

Throw away for now the first a_1 columns and the first row and consider the remaining $(1000 - a_1) \times 999$ rectangular grid G_2. It contains $1999 - a_1 \geq 999 + 2 = 1001$ colored squares. Therefore, there is a row in G_2 with at least $a_2 \geq 2$ colored squares. Interchange rows and then columns so that the first a_2 squares of the first row are colored. Then the first a_2 columns have no colored squares other than the ones in the first row.

Observe that $a_1 + a_2$ cannot equal to 1000, for then we would have no place to put the remaining 999 colored squares. Also, $a_1 + a_2$ cannot equal 999, for then the remaining 1000 colored squares must all be in the last column and we would have a right triangle, a contradiction. Hence $1000 - (a_1 + a_2) \geq 2$.

The above process can be continued, but $1000 - (a_1 + a_2 + \cdots) \geq 2$ contradicts the fact that $a_1, a_2, \cdots \geq 2$. Thus, with 1999 colored squares there must be a right triangle. As in the first solution, we can find a way to arrange 1998 colored squares without obtaining a right triangle of the type described.

Third Solution. We prove a more general statement:

Lemma. Let $n_{k,\ell}$ be the smallest positive integer such that if $n_{k,\ell}$ squares of a $k \times \ell$ ($k, \ell \geq 2$) board are colored, then there necessarily exist a right triangle of the type described. Define $t = t_{k,\ell} = k + \ell$ to be the total dimension of the board. Then $n_{k,\ell} = t - 1$.

Proof. As in the first solution, we can color $t-2$ squares without producing a right triangle: fill every square in the first row and the first column, except for the one square at their intersection. Hence $n_{k,\ell} \geq t-1$.

Now we prove by induction on t that $n_{k,\ell} = t-1$.

For the base case $t = 4$, we have $k = \ell = 2$ and it is easy to see that $n_{2,2} = 3$.

Assume that the claim is true for $t = m$, $m \geq 4$. For $t = m+1$, we claim that $n_{k,\ell} = m$ when $k+\ell = m+1$, $k,\ell \geq 2$. For the sake of contradiction, suppose that there is a $k \times \ell$ board with m colored squares and no right triangles. Without loss of generality, suppose that $k \geq \ell$. Then $k > 3$. There is a row with at most 1 colored square because otherwise we will have at least $2k \geq t > m$ colored squares. Cross out that row to obtain a $(k-1) \times \ell$ board with $t = m$, and $k-1, \ell \geq 2$, and at least $\geq m-1$ colored squares. By the induction hypothesis, there is right triangle, contradicting our assumption. Therefore our assumption is wrong and we conclude that $n_{k,\ell} = m = t-1$. Our induction is complete, and this finishes our proof. □

5. Let $A_1A_2A_3$ be a triangle and let ω_1 be a circle in its plane passing through A_1 and A_2. Suppose there exist circles $\omega_2, \omega_3, \ldots, \omega_7$ such that for $k = 2, 3, \ldots, 7$, ω_k is externally tangent to ω_{k-1} and passes through A_k and A_{k+1}, where $A_{n+3} = A_n$ for all $n \geq 1$. Prove that $\omega_7 = \omega_1$.

First Solution.

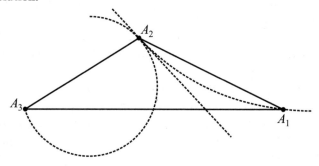

Without loss of generality we may assume that in counterclockwise order, the vertices of the triangle are A_1, A_2, A_3. Let θ_1 be the measure of the arc from A_1 to A_2 along ω_1, taken in the counterclockwise direction. Define $\theta_2, \ldots, \theta_7$ analogously.

Let ℓ be the line through A_2 tangent to ω_1 and ω_2. Then the angle from the line A_1A_2 to ℓ, again measured counterclockwise, is $\theta_1/2$. Similarly, the angle from ℓ to A_2A_3 is $\theta_2/2$. Therefore, writing $\angle A_1A_2A_3$ for the counterclockwise angle from the line A_1A_2 to the line A_2A_3, we have

$$\theta_1 + \theta_2 = 2\angle A_1A_2A_3.$$

By similar reasoning we obtain the system of six equations:

$$\theta_1 + \theta_2 = 2\angle A_1A_2A_3, \theta_2 + \theta_3 = 2\angle A_2A_3A_1,$$
$$\theta_3 + \theta_4 = 2\angle A_3A_1A_2, \theta_4 + \theta_5 = 2\angle A_1A_2A_3$$
$$\theta_5 + \theta_6 = 2\angle A_2A_3A_1, \theta_6 + \theta_7 = 2\angle A_3A_1A_2.$$

Adding the equations on the left column, and subtracting the equations on the right yields $\theta_1 = \theta_7$.

To see that this last equality implies $\omega_1 = \omega_7$, simply note that as the center O of a circle passing through A_1 and A_2 moves along the perpendicular bisector of A_1A_2, the angle θ_1 goes monotonically from 0 to 2π. Thus the angle determines the circle.

Second Solution.

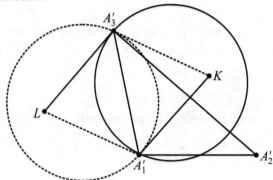

Perform an **inversion** with center A_2 and arbitrary radius. For an object P, let P' denote the image of P under the **inversion**. We may assume that A'_2 is the point at infinity. Then ω'_1 is a line through A'_1, and ω'_2 is the line parallel to ω'_1 passing through A'_3. Next, ω'_3 will be the circle passing through A'_1 tangent to ω'_2 at A'_3 and ω'_4 will be the line tangent to ω'_3 at A'_1. Let ω'_4 and ω'_2 meet at K. Since ω'_2 and ω'_4 are the tangents to ω'_3 through K, we have $KA'_1 = KA'_3$. Next, ω'_5 will be the line passing through A'_3 parallel to ω'_4. Let ω'_5 and ω'_1 meet at L. Note that $A'_1KA'_3L$ is a rhombus, since opposite sides are parallel and $KA'_1 = KA'_3$. Next, ω'_6 is the circle through A'_1 and tangent to ω'_5 at A'_3. Since $LA_1 = LA_3$,

ω'_6 is also tangent to LA'_1 (which is ω'_1) at A'_1. Finally, ω'_7 will be the line tangent to ω'_6 at A'_1. But by the remarks above this is ω'_1. Hence $\omega'_1 = \omega'_7$ and, thus, $\omega_1 = \omega_7$.

6. Let $a_1, b_1, a_2, b_2, \ldots, a_n, b_n$ be nonnegative real numbers. Prove that
$$\sum_{i,j=1}^{n} \min\{a_i a_j, b_i b_j\} \leq \sum_{i,j=1}^{n} \min\{a_i b_j, a_j b_i\}.$$

Comment. If you were not able to make significant progress on this problem, you are not alone. There was no complete solution submitted by USAMO contestants.

Solution. (Based on work by George Lee) Define
$$L(a_1, b_1, \ldots, a_n, b_n) = \sum_{i,j} (\min\{a_i b_j, a_j b_i\} - \min\{a_i a_j, b_i b_j\}).$$

Our goal is to show that
$$L(a_1, b_1, \ldots, a_n, b_n) \geq 0$$
for $a_1, b_1, \ldots, a_n, b_n \geq 0$. Our proof is by induction on n, the case $n = 1$ being evident. Using the obvious identities

- $L(a_1, 0, a_2, b_2, \ldots) = L(0, b_1, a_2, b_2, \ldots) = L(a_2, b_2, \ldots)$,
- $L(x, x, a_2, b_2, \ldots) = L(a_2, b_2, \ldots)$,

and the less obvious but easily verified identities

- $L(a_1, b_1, a_2, b_2, a_3, b_3, \ldots) = L(a_1 + a_2, b_1 + b_2, a_3, b_3, \ldots)$ if $a_1/b_1 = a_2/b_2$,
- $L(a_1, b_1, a_2, b_2, a_3, b_3, \ldots) = L(a_2 - b_1, b_2 - a_1, a_3, b_3, \ldots)$ if $a_1/b_1 = b_2/a_2$ and $a_1 \leq b_2$,

we may deduce the result from the induction hypothesis unless we are in the following situation:

1. all of the a_i and b_i are nonzero;
2. for $i = 1, \ldots, n$, $a_i \neq b_i$;
3. for $i \neq j$, $a_i/b_i \neq a_j/b_j$ and $a_i/b_i \neq b_j/a_j$.

For $i = 1, \ldots, n$, let $r_i = \max\{a_i/b_i, b_i/a_i\}$. Without loss of generality, we may assume $1 < r_1 < \cdots < r_n$, and that $a_1 < b_1$. Now notice that $f(x) = L(a_1, x, a_2, b_2, \ldots, a_n, b_n)$ is a *linear* function of x in the interval

$[a_1, r_2 a_1]$. Explicitly,

$$f(x) = \min\{a_1 x, xa_1\} - \min\{a_1^2, x^2\} + L(a_2, b_2, \ldots, a_n, b_n)$$
$$+ 2\sum_{j=2}^{n}(\min\{a_1 b_j, xa_j\} - \min\{a_1 a_j, xb_j\})$$
$$= (x - a_1)(a_1 + 2\sum_{j=2}^{n} c_j) + L(a_2, b_2, \ldots, a_n, b_n),$$

where $c_j = -b_j$ if $a_j > b_j$ and $c_j = a_j$ if $a_j < b_j$.

In particular, since f is linear, we have

$$f(x) \geq \min\{f(a_1), f(r_2 a_1)\}.$$

Note that $f(a_1) = L(a_1, a_1, a_2, b_2, \ldots) = L(a_2, b_2, \ldots)$ and

$$f(r_2 a_1) = L(a_1, r_2 a_1, a_2, b_2, \ldots)$$
$$= \begin{cases} L(a_1 + a_2, r_2 a_1 + b_2, a_3, b_3, \ldots) & \text{if } r_2 = b_2/a_2, \\ L(a_2 - r_2 a_1, b_2 - a_1, a_3, b_3, \ldots) & \text{if } r_2 = a_2/b_2. \end{cases}$$

Thus we deduce the desired inequality from the induction hypothesis in all cases.

Note. More precisely, it can be shown that for $a_i, b_i > 0$, equality holds if and only if, for each $r > 1$, the set S_r of indices i in $\{1, \ldots, n\}$ such that $a_i/b_i \in \{r, 1/r\}$ has the property that

$$\sum_{i \in S_r} a_i = \sum_{i \in S_r} b_i.$$

Namely, assume this is the case for $n - 1$ pairs. Given n pairs, if conditions (a)–(c) are not all met, we may deduce the result from the induction hypothesis by the same reductions as that at the beginning of the proof. If (a)–(c) are met, then for equality to hold, $0 = f(b_1) \geq \min\{f(a_1), f(r_2 a_1)\} \geq 0$. Since $f(x)$ is linear on the interval $[a_1, r_2 a_1]$, $f(x)$ is identically zero on the interval. Since $f(a_1) = 0$, $L(a_2, b_2, \ldots) = 0$. Applying the induction hypothesis, with all of a_i and b_i nonzero ((a) is met) and $r_i > 1$ ((b) is met), we have

(i) for each $i \geq 2$, there exists an $r > 1$ such that either $a_i = rb_i$ or $b_i = ra_i$.

(ii) $\sum_{j=2}^{n}(a_j - b_j) = 0.$

Therefore, if $a_j > b_j$, $a_j = rb_j$ and $c_j(1-r) = (-b_j)(1-r) = a_j - b_j$; if $a_j < b_j$, $ra_j = b_j$ and $c_j(1-r) = a_j(1-r) = a_j - b_j$. Hence $\sum_{j=2}^n c_j = 0$. But then $0 = f(r_2 a_1)$ yields

$$0 = (r_2 a_1 - a_1)(a_1 + 2\sum_{j=2}^n c_j) + L(a_2, b_2, \ldots, a_n, b_n) = (r_2 - 1)a_1^2$$

and $a_1 = 0$, a contradiction.

2 Team Selection Test

1. Let a, b, c be nonnegative real numbers. Prove that
$$\frac{a+b+c}{3} - \sqrt[3]{abc} \le \max\{(\sqrt{a}-\sqrt{b})^2, (\sqrt{b}-\sqrt{c})^2, (\sqrt{c}-\sqrt{a})^2\}.$$

First Solution. We prove the stronger inequality
$$a+b+c - 3\sqrt[3]{abc} \le (\sqrt{a}-\sqrt{b})^2 + (\sqrt{b}-\sqrt{c})^2 + (\sqrt{c}-\sqrt{a})^2. \quad (1)$$

The conclusion is immediate if $abc = 0$, so we assume that $a, b, c > 0$. By multiplying a, b, c by a suitable factor, we may reduce to the case $abc = 1$. Without loss of generality, assume that a and b are both greater than or equal to 1, or both less than or equal to 1. The desired inequality now becomes

$$0 \le a+b+c - 2\sqrt{ab} - 2\sqrt{bc} - 2\sqrt{ca} + 3$$
$$= (\sqrt{a}-\sqrt{b})^2 + \frac{1}{ab} - \frac{2}{\sqrt{a}} - \frac{2}{\sqrt{b}} + 3$$
$$= (\sqrt{a}-\sqrt{b})^2 + \left(\frac{1}{\sqrt{a}}-1\right)^2 + \left(\frac{1}{\sqrt{b}}-1\right)^2 + \frac{1}{ab} - \frac{1}{a} - \frac{1}{b} + 1$$
$$= (\sqrt{a}-\sqrt{b})^2 + \left(\frac{1}{\sqrt{a}}-1\right)^2 + \left(\frac{1}{\sqrt{b}}-1\right)^2 + \left(\frac{1}{a}-1\right)\left(\frac{1}{b}-1\right).$$

Second Solution. (by Ian Le) We again prove the stronger inequality (1), which can be rewritten

$$\sum_{\text{sym}} \left[a - 2(ab)^{1/2} + (abc)^{1/3}\right] \ge 0,$$

where the sum is taken over all six permutations of a, b, c. This inequality follows from adding the two inequalities

$$\sum_{\text{sym}} [a - 2a^{2/3}b^{1/3} + (abc)^{1/3}] \ge 0$$

and

$$\sum_{\text{sym}} [a^{2/3}b^{1/3} + a^{1/3}b^{2/3} - 2a^{1/2}b^{1/2}] \ge 0.$$

The first of these is **Schur's inequality** with $x = a^{1/3}, y = b^{1/3}, z = c^{1/3}$, while the second follows from the **AM-GM inequality**.

Third Solution. Without loss of generality, assume that b is between a and c. The desired inequality reads

$$a + b + c - 3\sqrt[3]{abc} \leq 3(c + a - 2\sqrt{ac}).$$

As a function of b, the right side minus the left side is concave (its second derivative is $-(2/3)(ac)^{1/3}b^{-5/3}$), so its minimum value in the range $[a, c]$ occurs at one of the endpoints. Thus, without loss of generality, we may assume $a = b$. Moreover, we may rescale the variables to get $a = b = 1$. Now the claim reads

$$\frac{2c + 3c^{1/3} + 1}{6} \geq c^{1/2}.$$

This is an instance of a weighted **AM-GM inequality**.

Note. More generally, for nonnegative real numbers $a_1, a_2, \ldots, a_n$, we have

$$\frac{m}{2} \leq \frac{a_1 + a_2 + \cdots + a_n}{n} - \sqrt[n]{a_1 a_2 \cdots a_n} \leq \frac{(n-1)M}{2}, \quad (2)$$

where

$$m = \min_{1 \leq i < j \leq n} \{(\sqrt{a_i} - \sqrt{a_j})^2\} \text{ and } M = \max_{1 \leq i < j \leq n} \{(\sqrt{a_i} - \sqrt{a_j})^2\}.$$

The right inequality can be proved, by using the method of the third solution above. We leave the details as an exercise for the reader.

The left inequality falls apart when we replace m by c, the average of $(\sqrt{a_i} - \sqrt{a_j})^2$ for $1 \leq i < j \leq n$. Since

$$\frac{m}{2} \leq \frac{c}{2} = \frac{\sum_{1 \leq i < j \leq n}(\sqrt{a_i} - \sqrt{a_j})^2}{2\binom{n}{2}}$$

$$= \frac{\sum_{1 \leq i < j \leq n}(\sqrt{a_i} - \sqrt{a_j})^2}{n(n-1)}$$

$$= \frac{(n-1)(a_1 + a_2 + \cdots + a_n) - 2\sum_{1 \leq i < j \leq n}\sqrt{a_i a_j}}{n(n-1)},$$

the inequality now reads

$$\sum_{1 \leq i < j \leq n} \sqrt{a_i a_j} \geq \frac{n(n-1)\sqrt[n]{a_1 a_2 \cdots a_n}}{2}.$$

This follows from the **AM-GM inequality**.

We may also replace m by

$$m' = \min_{1 \le k \le n} \{(\sqrt{a_k} - \sqrt{a_{k+1}})^2\}$$

in (2) to obtain, in a way, a sharper lower bound. A similar proof works. We leave it to the reader as an exercise.

Even more generally, one can ask for a comparison between the difference between the arithmetic and geometric means of a set of n nonnegative real numbers, and the maximum (or average) difference between the arithmetic and geometric means over all k-element subsets. The authors do not know what the correct inequalities should look like or how they may be proved.

2. Let $ABCD$ be a cyclic quadrilateral and let E and F be the feet of perpendiculars from the intersection of diagonals AC and BD to $\overline{AB}$ and $\overline{CD}$, respectively. Prove that $\overline{EF}$ is perpendicular to the line through the midpoints of $\overline{AD}$ and $\overline{BC}$.

First Solution.

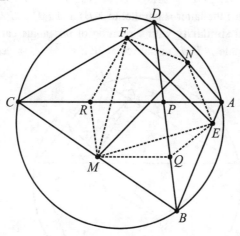

Let M and N be the midpoints of $\overline{BC}$ and $\overline{DA}$, respectively. We will show that $EM = FM$ and $EN = FN$, from which it will follow that E and F are reflections of each other across the line MN, and in particular $\overline{EF} \perp \overline{MN}$.

Let P be the intersection of $\overline{AC}$ and $\overline{BD}$, and Q and R the midpoints of $\overline{BP}$ and $\overline{CP}$, respectively. Then $EQ = PQ$, because $\overline{EQ}$ is a median of the right triangle PEB, and $PQ = RM$, because $PQMR$

is a parallelogram. Similarly, $FR = PR = QM$. Moreover, with the angles being directed $\mod 180°$, we have

$$\begin{aligned}\angle EQM &= \angle EQP + \angle PQM \\ &= 2\angle EBP + \angle PQM \\ &= \angle MRP + 2\angle PCF \\ &= \angle MRP + \angle PRF = \angle MRF,\end{aligned}$$

where we have used the equalities $\angle ABD = \angle ACD$ from the cyclic quadrilateral $ABCD$, and $\angle PQM = \angle MRQ$ from the parallelogram $PQMR$.

Putting this all together, we find that triangles EQM and MRF are congruent, so $EM = FM$. Similarly, $EN = FN$, and so $\overline{EF} \perp \overline{MN}$, as desired.

Second Solution. Keep the notation of the first solution, and also let T be the midpoint of $\overline{BD}$. Observe that triangles ABP and DCP are similar, so that $EP/FP = AB/CD = NT/TM$. Moreover, note that $\overline{EP} \perp \overline{NT}$ (since the latter is parallel to $\overline{AB}$) and $\overline{FP} \perp \overline{TM}$. Therefore, there is a **spiral similarity** with a rotation of $90°$ angle carrying triangle EPF onto triangle NTM; in particular, $\overline{EF} \perp \overline{MN}$, as desired.

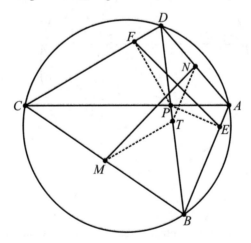

Third Solution. Keep the notation of the first solution, and introduce vectors using P as the origin. Then $M = (B+C)/2$, $N = (D+A)/2$, so $\overline{EF} \perp \overline{MN}$ is equivalent to the vector equalities

$$0 = (E - F) \cdot (M - N),$$
$$0 = (E - F) \cdot (B + C - D - A),$$
$$0 = E \cdot (B + C - D - A) - F \cdot (B + C - D - A),$$

and
$$0 = E \cdot (C - D) - F \cdot (B - A),$$

since $E \cdot (B - A) = F \cdot (D - C) = 0$ by the construction of E and F. This last equality can be rewritten as

$$\frac{PE}{AB} \cos \angle(PE, DC) = \frac{PF}{CD} \cos \angle(PF, AB).$$

Since triangles PAB and PDC are similar, and E and F are the feet of the altitudes from P, we have $PE/AB = PF/CD$. Moreover, the two (directed) angles are each equal to 90° plus the angle between lines AB and CD, and so are equal to each other.

3. For real number x, let $\lceil x \rceil$ denote the smallest integer greater than or equal to x, let $\lfloor x \rfloor$ denote the greatest integer less than or equal to x, and let $\{x\}$ denote the fractional part of x, which is given by $x - \lfloor x \rfloor$. Let p be a prime number. For integers r, s such that $rs(r^2 - s^2)$ is not divisible by p, let $f(r, s)$ denote the number of integers $n \in \{1, 2, \ldots, p-1\}$ such that $\{rn/p\}$ and $\{sn/p\}$ are either both less than $1/2$ or both greater than $1/2$. Prove that there exists $N > 0$ such that for $p \geq N$ and all r, s,

$$\left\lceil \frac{p-1}{3} \right\rceil \leq f(r, s) \leq \left\lfloor \frac{2(p-1)}{3} \right\rfloor.$$

Solution. We assume that p is sufficiently large. Since $f(r, s) = f(br, bs)$ for any b not divisible by p, we may assume $r = 1$ and simply write $f(s)$ instead of $f(r, s)$. Also notice that $f(s) = p - 1 - f(-s)$, so we may assume $1 \leq s \leq (p-1)/2$. Moreover, $s = 1$ is forbidden, $f(2) = (p-1)/2$ or $(p-3)/2$, and one easily checks that $f(3) = \lfloor 2(p-1)/3 \rfloor$. So we may assume $s \geq 4$.

Let $g_1(s)$ be the number of $a \in \{1, \ldots, (p-1)/2\}$ such that $\{as/p\} < 1/2$; and let $g_2(s)$ be the number of $a \in \{(p+1)/2, \ldots, p-1\}$ such that $\{as/p\} > 1/2$. Note that

$$\{as/p\} + \{(p-a)s/p\} = 1.$$

Thus $g_1(s) = g_2(s) = f(s)/2$. We consider two cases.

1. $s \in [4, p/4]$. Note that
$$\left| f(s) - \frac{p-1}{2} \right| = \left| g_1(s) - \left(\frac{p-1}{2} - g_1(s) \right) \right| \le \frac{s}{2} + \frac{p}{2s}.$$
To see this, divide $s, 2s, \ldots, (p-1)s/2$ into groups depending on which of the intervals $(0, p), (p, 2p), \ldots$ they fall into. In each group except the last one, at most one more number falls into one half of the interval than into the other half. Since the largest of these numbers is less than $sp/2$, the number of such groups is at most $s/2$. In the last group, the maximum discrepancy is the greatest quantity of the numbers that fit into one half of the interval, which is at most $p/2s$.

For $s \in [4, p/4]$, the right side of the above inequality achieves its maximum value at the endpoints of the interval. Thus we get the upper bound
$$\left| f(r, s) - \frac{p-1}{2} \right| \le 2 + \frac{p}{8}$$
and the right side is less than $(p-1)/6$ for p sufficiently large.

2. $s \in ((p-1)/4, (p-1)/2]$. In this case, there cannot exist three consecutive values of $a \in \{1, \ldots, (p-1)/2\}$ or three consecutive values of $a \in \{(p+1)/2, \ldots, p-1\}$ such that the three values of $\{as/p\}$ all lie in a single interval of length $1/2$. For $p \equiv 1 \pmod{6}$ this implies $\lceil (p-1)/3 \rceil \le f(s) \le \lfloor 2(p-1)/3 \rfloor$; for $p \equiv 5 \pmod{6}$, the lower bound is true. To violate the upper bound, $f(s) \ge 4k + 3$ where $p = 6k + 5$. Since $f(s) = 2g_2(s)$, $g_2(s) \ge 2k + 2$. But we can regroup $3k + 2$ numbers $\{1, 2, \ldots, (p-1)/2\}$ as
$$\underbrace{\{(1,2), (3,4,5), \ldots, (3k, 3k+1, 3k+2)\}}_{k+1 \text{ groups}}.$$

From the earlier observation, each group can provide at most two a's such that $\{as/p\} < 1/2$. Hence $\{s/p\}, \{2s/p\} < 1/2$. Since $1 \le s \le (p-1)/2$, $\{2s/p\} = 2s/p$. But then $s/p < 1/4$ and $4s < p$. Now $p > 4s > p - 1$, which is impossible for integers s and p.

Note. Paul Valiant noted that one can alternatively treat the case $s > (p-1)/4$ by a more careful analysis of the group sizes, particularly in the neighborhood of $(p-1)/3$.

The assertion of the problem holds in fact for all $p \ge 5$ (note that the assertion is vacuous for $p = 2, 3$); furthermore, equality holds if and only if $r \equiv \pm 3s \pmod{p}$ or $s \equiv \pm 3r \pmod{p}$.

The result is the main step in the solution of the following problem, posed recently by Greg Martin. Fix a prime number p. I choose 3 integers a_1, a_2, a_3 not divisible by p and no two congruent modulo p. You then choose an integer r not divisible by p, and then collect from me a number of dollars equal to the smallest positive integer congruent to one of ra_1, ra_2, ra_3 modulo p. What is the smallest amount I will have to pay out, and how do I achieve this minimum? (The corresponding question for k integers instead of 3 is open, and the proposer offers \$15 for its solution.)

4. Let n be a positive integer. Prove that

$$\sum_{i=0}^{n} \binom{n}{i}^{-1} = \frac{n+1}{2^{n+1}} \sum_{i=1}^{n+1} \frac{2^i}{i}.$$

Solution. Let

$$S_n = \frac{1}{n+1} \sum_{k=0}^{n} \binom{n}{k}^{-1} = \sum_{k=0}^{n} \frac{k!(n-k)!}{(n+1)!}.$$

We must show that $S_n = (\sum_{k=1}^{n+1} 2^k/k)/2^{n+1}$. To do so, it suffices to check that $S_1 = 1$, which is clear, and that $2^{n+2} S_{n+1} - 2^{n+1} S_n = 2^{n+2}/(n+2)$. Now

$$2 S_{n+1} = \frac{1}{n+2} \left(\sum_{i=0}^{n+1} \binom{n+1}{i}^{-1} + \sum_{j=0}^{n+1} \binom{n+1}{j}^{-1} \right)$$

$$= \frac{2}{n+2} + \frac{1}{n+2} \sum_{i=0}^{n} \left(\binom{n+1}{i}^{-1} + \binom{n+1}{i+1}^{-1} \right)$$

$$= \frac{2}{n+2} + \frac{1}{n+2} \sum_{i=0}^{n} \frac{i!(n+1-i)! + (i+1)!(n-i)!}{(n+1)!}$$

$$= \frac{2}{n+2} + \frac{1}{n+2} \sum_{i=0}^{n} \frac{i!(n-i)!(n+1-i+i+1)}{(n+1)!}$$

$$= S_n + \frac{2}{n+2},$$

as claimed.

5. Let n be a positive integer. A *corner* is a finite set C of ordered n-tuples of positive integers such that if $a_1, a_2, \ldots, a_n, b_1, b_2, \ldots, b_n$ are positive integers with $a_k \geq b_k$ for $k = 1, 2, \ldots, n$ and $(a_1, a_2, \ldots, a_n) \in C$, then $(b_1, b_2, \ldots, b_n) \in C$. Prove that among any infinite collection S of corners, there exist two corners, one of which is a subset of the other one.

Comment. Working on parallel arguments for $n = 2$ and drawing pictures will help those readers not familiar with n-dimensional arguments.

First Solution. We prove that for any sequence $A_1, A_2, \ldots$ of corners (not necessarily distinct), there exists an increasing sequence $i_1, i_2, \ldots$ of integers such that $A_{i_1} \subseteq A_{i_2} \subseteq \cdots$. We do so by induction on n, the case $n = 1$ being obvious.

Define a *slice* to be the set of n-tuples of the form

$$\{(x_1, \ldots, x_{k-1}, y, x_{k+1}, \ldots, x_n) : x_1, \ldots, x_n \in \mathbb{N}\}$$

for some $k \in \{1, \ldots, n\}$ and some y.

Let $i_1 = 1$; if possible, pick i_2 so that $A_{i_1} \subseteq A_{i_2}$, pick i_3 so that $A_{i_2} \subseteq A_{i_3}$, and so on. If all of the i_k can be defined in this fashion, we are done. So assume, on the contrary, that some A_{i_k} is not contained in A_j for any $j > i_k$. Then for all $j > i_k$, A_j is contained in the union of the slices having nonempty intersection with A_{i_k}: any corner not in the union contains an element $(x_1, \ldots, x_n)$ such that $x_i > y_i$ for each $(y_1, \ldots, y_n) \in A_{i_k}$, and so contains A_{i_k}.

Let $L_1, L_2, \ldots, L_m$ be the slices that meet A_{i_k}. By the induction hypothesis, the sequence $A_{i_k+1}, A_{i_k+2}, \ldots$ contains a subsequence whose intersections with L_1 are each contained in the next. That subsequence in turn contains a subsequence whose intersections with L_2 are each contained in the next, and so on. In particular, there is a subsequence $A_{j_1}, A_{j_2}, \ldots$ of $A_{i_k+1}, A_{i_k+2}, \ldots$ whose intersections with each of $L_1, \ldots, L_m$ are each contained in the next. But all of these corners are contained in $L_1 \cup \cdots \cup L_m$ as noted above, so in fact $A_{j_1} \subseteq A_{j_2} \subseteq \cdots$ and $j_1, j_2, \ldots$ satisfy the desired conclusion.

Second Solution. (by Reid Barton) If $a = (a_1, \ldots, a_n)$ and $b = (b_1, \ldots, b_n)$, we write $a \leq b$ if $a_i \leq b_i$ for $i = 1, \ldots, n$. We first note that every sequence of n-tuples of positive integers contains a subsequence which is nondecreasing with this definition. For $n = 1$, we may simply pick the smallest term, then the smallest term that comes later in the

Formal Solutions

sequence, and so on. For general n, first pick a subsequence which is nondecreasing in its first coordinate, then pick a subsequence of that which is also nondecreasing in its second coordinate, and so on.

For the sake of contradiction, suppose that there are no corners $A, B \in S$ with $A \subset B$. Let C_1 be a corner in S; then each other corner in S fails to contain one of the n-tuples in C_1. Since S is infinite and C_1 is finite, there exists an n-tuple a^1 in C_1 such that the set S_1 of corners not containing a^1 is infinite.

Now let C_2 be a corner in S_1; again, we may find an n-tuple a^2 in C_2 such that the set S_2 of corners not containing a^2 is infinite. Analogously, we recursively construct sequences C_k of corners, S_k of infinite sets of corners and a^k of n-tuples such that C_k is a corner in S_{k-1}, a^k is an n-tuple in C_k and S_k is the set of corners in S_{k-1} not containing a^k.

To conclude, simply notice that $a^i \not\leq a^j$ for $i \leq j$, since a^j is an element of a corner which does not contain a^i. This contradicts the result of the first paragraph.

Note. The assertion of the problem still holds if corners are not required to be finite; this statement is a recent theorem of Diane Maclagan, which turns out to yield a nontrivial result in algebraic geometry.

6. Let ABC be a triangle inscribed in a circle of radius R, and let P be a point in the interior of ABC. Prove that
$$\frac{PA}{BC^2} + \frac{PB}{CA^2} + \frac{PC}{AB^2} \geq \frac{1}{R}.$$

Solution.

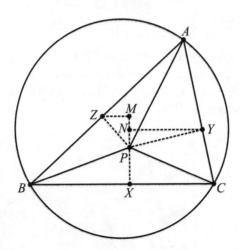

Let a, b, c, A, B, C be the sides and angles of triangle ABC. Let X, Y, Z be the feet of the perpendiculars from P to lines BC, CA, AB, respectively. Recall the inequality (the key ingredient in the proof of the **Erdős-Mordell inequality**):

$$PA \sin A \geq PY \sin C + PZ \sin B. \tag{1}$$

This says that the length of YZ is greater than equal to its projection onto $\overline{BC}$, the latter being equal to the sum of the lengths of the projections of $\overline{PY}$ and $\overline{PZ}$ onto $\overline{BC}$. In fact, since $\angle AYP = \angle AZP = 90°$, $AZPY$ is cyclic with $\overline{AP}$ as a diameter of its circumcircle. By the **Extended Law of Sines**, $YZ = PA \sin A$. Let M and N be the feet of perpendiculars from Z and Y to the line PX. Since $\angle BZP = \angle BXP = 90°$, $PZBX$ is cyclic. Hence $\angle MPZ = \angle B$ and $ZM = PZ \sin B$. Similarly, $YN = PY \sin C$. Thus (1) is equivalent to $YZ \geq YN + MZ$. Multiplying by $2R$ and using the Extended Law of Sines, (1) becomes

$$aPA \geq cPY + bPZ.$$

Likewise, we have $bPB \geq aPZ + cPX$ and $cPC \geq bPX + aPY$. Using these inequalities, we obtain

$$\frac{PA}{a^2} + \frac{PB}{b^2} + \frac{PC}{c^2}$$

$$\geq PX\left(\frac{b}{c^3} + \frac{c}{b^3}\right) + PY\left(\frac{c}{a^3} + \frac{a}{c^3}\right) + PZ\left(\frac{a}{b^3} + \frac{b}{a^3}\right)$$

$$\geq \frac{2PX}{bc} + \frac{2PY}{ca} + \frac{2PZ}{ab} \qquad \text{(AM-GM inequality)}$$

$$= \frac{4[ABC]}{abc} = \frac{1}{R}.$$

Equality in the first step requires that $\overline{YZ}$ be parallel to $\overline{BC}$ and so on. This occurs if and only if P is the circumcenter of ABC. Equality in the second step requires that $a = b = c$. Thus equality holds if and only if ABC is equilateral and P is its center.

3   IMO

1. Two circles ω_1 and ω_2 intersect at M and N. Line ℓ is tangent to the circles at A and B, respectively, so that M lies closer to ℓ than does N. Line CD, with C on ω_1 and D on ω_2, is parallel to ℓ and passes through M. Let lines AC and BD meet at E; let lines AN and CD meet at P; and let lines BN and CD meet at Q. Prove that $EP = EQ$.

First Solution.

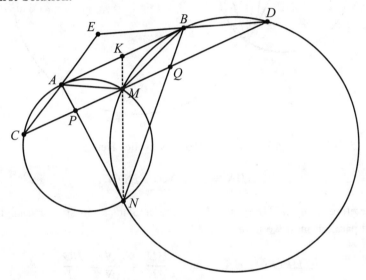

Let lines AB and MN meet at K. By the **Power of a point theorem**, $AK^2 = KN \cdot KM = BK^2$. Since $\overline{AB} \parallel \overline{PQ}$, $PM = QM$. Hence it suffices to prove that $\overline{EM} \perp \overline{PQ}$.

Since $\overline{CD} \parallel \overline{AB}$, $\angle EAB = \angle ECM$. Since $\overline{AB}$ is tangent to the circle at A, $\angle BAM = \widehat{AM}/2 = \angle ACM$. Therefore $\angle EAB = \angle BAM$. Similarly $\angle EBA = \angle ABM$ and $\overline{AB}$ bisects both $\angle EAM$ and $\angle EBM$. Hence $AEBM$ is a **kite** and $\overline{EM} \perp AB$. Since $\overline{PQ} \parallel \overline{AB}$, we obtain $\overline{EM} \perp \overline{PQ}$, as desired.

Second Solution. (By Xingyi Yuan, China) Since $DBMN$ and $CAMN$ are cyclic, $\angle DBN = \angle DMN$ and $\angle CAN = \angle CMN$. Hence

$$\angle DBN + \angle CAN = \angle DMN + \angle CMN = 180°,$$

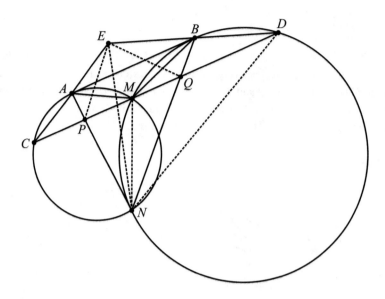

or $\angle DBN = \angle EAN$, that is, $AEBN$ is cyclic. Hence

$$\angle AEN = \angle ABN = \angle BDN.$$

Therefore triangles DBN and EAN are similar. From similar triangles and parallel lines, we have

$$\angle DBQ = \angle EAP \quad \text{and} \quad \frac{BD}{AE} = \frac{BN}{AN} = \frac{BQ}{AP}.$$

Thus triangles AEP and BDQ are similar and $\angle AEP = \angle BDQ$. Similarly, $\angle ACP = \angle BEQ$. Therefore

$$\angle EPQ = \angle AEP + \angle ACP = \angle BDQ + \angle BEQ = \angle EQP$$

and $EP = EQ$, as desired.

Note. The first part of this solution shows that $\angle ANE = \angle BND$. Similarly, $\angle CNA = \angle ENB$. Therefore

$$\angle CNE = \angle CNA + \angle ANE = \angle ENB + \angle BND = \angle END,$$

that is, $\overline{NE}$ bisects $\angle CND$. An alternative version of the problem, posed to the IMO, asked for a proof of this result.

Third Solution. (by George Lee)

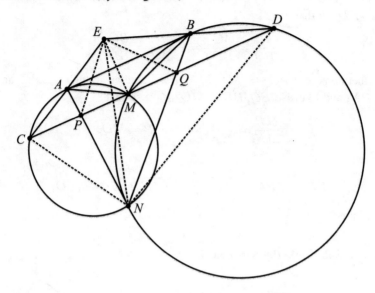

We know that $\angle DMB = \angle ABM$, because $AB \parallel DM$, and also that $\angle ABM = \angle BDM$ since $\angle ABM$ subtends $\widehat{BM}$ in ω_2. Thus, $\angle DMB = \angle BDM$, so triangle BDM is isosceles and B is the midpoint of $\widehat{MD}$. Similarly, A is the midpoint of $\widehat{MC}$. Then

$$\angle EBA = \angle BDM = \angle DMB = \angle ABM,$$

and similarly $\angle EAB = \angle BAM$. Thus triangles EBA and MBA are congruent, and M and E are reflections of each other across AB. Thus, $ME \perp AB$, $EA = MA = CA$, and $EB = MB = DB$.

Now note that

$$\angle ANB = \angle ANM + \angle MNB = \angle ACM + \angle MDB$$
$$= \angle ECD + \angle CDE = 180° - \angle DEC = 180° - \angle BEA,$$

so $EBNA$ is cyclic. Therefore

$$\angle CEN = \angle AEN = \angle ABN = \angle BDN = \angle EDN$$

and similarly $\angle NED = \angle NCE$, so triangles END and CNE are similar. But then AN and BN are corresponding medians of similar triangles END and CNE, so we have

$$\frac{AN}{BN} = \frac{EC}{ED}.$$

Also $AN/BN = AP/BQ$ since $AB \parallel PQ$, similarly $EC/ED = AC/BD$, so that

$$\frac{AP}{BQ} = \frac{AC}{BD} \iff \frac{AP}{AC} = \frac{BQ}{BD}.$$

Furthermore $AP/AC = MP/MN$ from similar triangles APC and MPN, and likewise $BQ/BD = MQ/MN$, so that

$$\frac{MP}{MN} = \frac{MQ}{MN} \iff MP = MQ.$$

Hence

$$EP = \sqrt{EM^2 + MP^2} = \sqrt{EM^2 + MQ^2} = EQ,$$

as desired.

Fourth Solution. (by Kiran Kedlaya)

Perform an **inversion** centered at M, and denote the image of a point by adding a prime to the name of the letter. Then the points C', D', N' form a triangle whose sides are formed by the inverses of the line CD and the circles ω_1 and ω_2. The inverse of the line ℓ is a circle tangent to lines $C'D'$, $D'N'$, $N'C'$ at M, B', A', respectively. Moreover, M lies between C and D, and so M also lies between C' and D'. We conclude that the inverse of ℓ is the excircle of $C'D'N'$ opposite N'. Let O be the center of this circle.

Now E' is the intersection of the circumcircles of triangles $A'MC'$ and $B'MD'$. Notice, though, that both of these circles pass through O. Thus $E' = O$; in particular, $E'M'$ is perpendicular to $C'D'$.

Note also that P' is the second intersection of the circumcircle of $A'N'M$ with the line $C'D'$; since $C'P' \cdot C'M = C'A' \cdot C'N'$ by the **Power of a point theorem**, and $C'M = C'A'$ by equal tangents, we have $C'P' = C'N'$, and analogously $D'Q' = D'N'$. By equal tangents $C'A' = C'M$, $D'B' = D'M$, and $A'N' = B'N'$,

$$MP' = MC' + C'P' = A'C' + C'N' = A'N'$$
$$= B'N' = B'D' + D'N' = MD' + D'Q' = MQ'.$$

We conclude that P' and Q' are reflections across the line $M'E'$, which implies that P and Q are reflections across the line ME. Thus $EP = EQ$, as desired.

2. Let a, b, c be positive real numbers such that $abc = 1$. Prove that

$$\left(a - 1 + \frac{1}{b}\right)\left(b - 1 + \frac{1}{c}\right)\left(c - 1 + \frac{1}{a}\right) \leq 1.$$

First Solution. Since $abc = 1$, this non-homogeneous inequality can be transformed into a homogeneous one by a suitable change of variables. In fact, there exist positive real numbers p, q, r such that

$$a = \frac{p}{q}, \ b = \frac{q}{r}, \ c = \frac{r}{p}.$$

Rewriting the inequality in terms of p, q, r, we obtain

$$(p - q + r)(q - r + p)(r - p + q) \leq pqr, \qquad (1)$$

where $p, q, r > 0$.

At most one of the numbers $u = p - q + r$, $v = q - r + p$, $w = r - p + q$ is negative, because any two of them have a positive sum. If exactly one of the numbers u, v, w is negative, then $uvw \leq 0 < pqr$. If they are all nonnegative, then by the **AM-GM inequality**,

$$\sqrt{uv} \leq \frac{1}{2}(u + v) = p.$$

Likewise, $\sqrt{vw} \leq q$ and $\sqrt{wu} \leq r$. Hence $uvw \leq pqr$, as desired.

Second Solution. Expanding out the left-hand side of (1) gives

$$(p - q + r)(q - r + p)(r - p + q)$$
$$= [p(p - r) + (r - q)(p - r) + q(r - q) + pq][r + (q - p)]$$
$$= pr(p - r) + r(r - q)(p - r) + rq(r - q) + pqr$$
$$+ p(p - r)(q - p) + (r - q)(p - r)(q - p)$$
$$+ q(r - q)(q - p) + pq(q - p).$$

Note that

$$pr(p - r) + rq(r - q) + pq(q - p) + (r - q)(p - r)(q - p) = 0.$$

Thus (1) is equivalent to

$$0 \le p(p - q)(p - r) + q(q - r)(q - p) + r(r - p)(r - q),$$

which is a special case of **Schur's inequality**.

Third Solution. Denoting the left-hand side of the desired inequality by L, we have

$$L = abcL = b\left(a - 1 + \frac{1}{b}\right) c\left(b - 1 + \frac{1}{c}\right) a\left(c - 1 + \frac{1}{a}\right)$$
$$= (ab - b + 1)(bc - c + 1)(ca - a + 1) = L_1.$$

Also since $1/b = ac$, $1/c = ab$, $1/a = bc$,

$$L = \left(a - 1 + \frac{1}{b}\right)\left(b - 1 + \frac{1}{c}\right)\left(c - 1 + \frac{1}{a}\right)$$
$$= (a - 1 + ac)(b - 1 + ab)(c - 1 + bc) = L_2.$$

If $u = a - 1 + 1/b \le 0$, then $a < 1$ and $b > 1$, implying that

$$v = b - 1 + 1/c > 0 \quad \text{and} \quad w = c - 1 + 1/a > 0.$$

Then $L = uvw \le 0$, as desired. Similarly, either $u \le 0$ or $v \le 0$ yields the same result. If $u, v, w > 0$, then all factors of L_1 and L_2 are positive. The **AM-GM inequality** gives

$$\sqrt{(ab - b + 1)(b - 1 + ab)} \le \frac{1}{2}[(ab - b + 1) + (b - 1 + ab)] = ab,$$

Likewise,

$$\sqrt{(bc - c + 1)(c - 1 + bc)} \le bc,$$

$$\sqrt{(ca-a+1)(a-1+ac)} \le ca.$$
Hence $L = \sqrt{L_1 L_2} \le (ab)(bc)(ca) = (abc)^2 = 1$.

Fourth Solution. Using the notations established in the third solution, it is easy to verify the equalities
$$bcu + cv = 2,$$
$$cav + aw = 2,$$
$$abw + bu = 2.$$

As in the third solution, we only need to consider the case when $u, v, w > 0$. The **AM-GM inequality** gives
$$2 \ge 2c\sqrt{buv},\ 2 \ge 2a\sqrt{cvw},\ \text{and}\ 2 \ge 2b\sqrt{awu},$$
from which $uvw \le 1$.

Fifth Solution. Let $u_1 = ab - b + 1$, $v_1 = bc - c + 1$, $w_1 = ca - a + 1$; $u_2 = 1 - bc + c$, $v_2 = 1 - ca + a$, and $w_2 = 1 - ab + b$. As in the third solution, we only need to consider the case in which $u_i, v_i, w_i > 0$ for $i = 1, 2$. Again, we have
$$L = u_1 v_1 w_1 = u_2 v_2 w_2.$$
Let $X = a + b + c$ and $Y = ab + bc + ca$. Then
$$u_1 + v_1 + w_1 = Y - X + 3 \text{ and } u_2 + v_2 + w_2 = X - Y + 3.$$
Hence either $u_1 + v_1 + w_1 \le 3$ or $u_2 + v_2 + w_2 \le 3$. In either case $L \le 1$ follows from the **AM-GM inequality**.

3. Let $n \ge 2$ be a positive integer. Initially, there are n fleas on a horizontal line, not all at the same point. For a positive real number λ, define a *move* as follows:

choose any two fleas, at points A and B, with A to the left of B; let the flea at A jump to the point C on the line to the right of B with $BC/AB = \lambda$.

Determine all values of λ such that, for any point M on the line and any initial positions of the n fleas, there is a finite sequence of moves that will take all the fleas to positions to the right of M.

Comment. This was the hardest problem on this IMO. There were only 15 complete solutions, and 3 of them were from the USA (by Reid Barton, George Lee, and Ricky Liu). Many students proved that $\lambda \geq 1/(n-1)$ is a sufficient condition (getting 2 out of 7 points), but failed to find a monovariant (functions that don't increase when a move is made) in order to prove that the condition $\lambda \geq 1/(n-1)$ is also necessary.

First Solution. (based on work by George Lee, Paul Valiant, Reid Barton, and the proposer) The answer is $\lambda \geq \dfrac{1}{n-1}$.

1. *If $\lambda \geq 1/(n-1)$, then the fleas can all move to the right of M.* We present two arguments.

First argument. Let the original leftmost flea be at $[0]$ and let $[x]$ be the point x to the right of $[0]$; and say the fleas can *conquer* a point if they can get arbitrarily close to, on, or to the right of that point.

Lemma. Suppose that we are given a fixed λ and $n \geq 2$ fleas on the line. If they cannot jump arbitrarily far to the right, then $\lambda < 1/(n-1)$, and the fleas can conquer the point $[k_n d_1]$, where

$$k_n = \frac{1 - (n-2)\lambda}{1 - (n-1)\lambda}$$

and $d_1 > 0$ is the distance between the first two distinct points containing fleas.

Proof. We prove the lemma by induction on n.

For $n = 2$, assume that two fleas cannot jump arbitrarily far to the right. After the fleas hop over each other t times the leftmost flea will be at

$$\left[d_1(1 + \lambda + \lambda^2 + \cdots + \lambda^{t-1})\right].$$

If $\lambda \geq 1$, the above geometric series diverges so that the fleas can jump arbitrarily far to the right, a contradiction. Hence $\lambda < 1$. Then the above geometric series converges to $[k_2 d_1]$, that is, the fleas can conquer this point.

Now suppose we have n fleas and the claim is true for $n - 1$ fleas. If $\lambda \geq 1/(n-2)$, then the first $n - 1$ fleas can jump arbitrarily far to the right and then the last flea can jump over them. In the other case, $\lambda < 1/(n-2)$. Note that this implies that $k = k_{n-1}$ is positive and greater than 1.

First, if there is more than one flea on the leftmost point, hop all but one over the rightmost flea so that only one flea remains. Now we perform a series of steps $1, 2, \ldots, t$. Fix some small $\epsilon > 0$ such that each value $d'_i = (k-1)^{i-1} d_1 - k^i \epsilon$ is positive for $i = 1, 2, \ldots, t$. In step i, label the fleas **A**, **B**, **C**, ... from left to right and suppose the distance between fleas **A** and **B** is $d \geq d'_i$. Then the distance between fleas **A** and **C** is at least d as well. If we ignore flea **B**, then by the induction hypothesis the other fleas can conquer the point kd to the right of **A**, or $(k-1)d$ to the right of **B**. Specifically, they can all move past the point $(k-1)d - k^i \epsilon \geq (k-1)^i d_1 - (k-1) k^i \epsilon - k^i \epsilon = d'_{i+1}$ to the right of **B**. Thus after t such steps the fleas can move past $[d'_1 + d'_2 + \cdots + d'_t]$, or

$$\left[d_1 \sum_{i=1}^{t} (k-1)^{i-1} - \epsilon \sum_{i=1}^{t} k^i \right].$$

Since ϵ can be made arbitrarily small, this implies that the fleas can conquer

$$\left[d_1 \left(1 + (k-1) + \cdots + (k-1)^{t-1} \right) \right],$$

which is a geometric series with ratio $k - 1$. If $\lambda \geq 1/(n-1)$, then this ratio is at least 1, so the geometric series diverges, and the fleas can conquer any point. If $\lambda < 1/(n-1)$ then the geometric series converges to

$$\left[d_1 \cdot \frac{1}{1 - (k_{n-1} - 1)} \right] = [k_n d_1],$$

so the fleas can conquer $[k_n d_1]$. This completes the induction. □

Second argument. Assume without loss of generality that the fleas are all at distinct points; otherwise we can attain such an arrangement by repeatedly jumping the leftmost flea at any given time over the rightmost flea at that time. Let k be the original minimum distance between any two adjacent fleas and let D be the original distance from the leftmost position to M, the point we wish to pass.

Originally, the leftmost flea **L** is at least $k(n-1)$ distance away from the rightmost flea **R**; have **L** jump over **R**. Then **L** will land at least

$$k(n-1) \cdot \lambda \geq k$$

to the right of **R**. Thus we have moved the left side of our flea circus at least k distance to the right, while keeping the minimum distance between any two adjacent fleas at least k. Then after at most $\lceil \frac{D}{k} \rceil$ moves of this sort, all the fleas will be to the right of M, as desired.

2. *If $\lambda < 1/(n-1)$, then the fleas cannot always all move to the right of M.*

As the fleas jump, let $O = (F_1, F_2, \ldots, F_n)$, where the fleas (from left to right) are at points $F_1, F_2, \ldots, F_n$. Then let

$$P(O) = F_1 F_n + F_2 F_n + \cdots + F_{n-1} F_n.$$

We claim that if any flea **F** jumps a distance d from its position, then P decreases by at least γd, where

$$\gamma = \frac{1 - (n-1)\lambda}{1 + \lambda}.$$

(Notice that γ is positive since $\lambda < 1/(n-1)$, and that $\gamma < 1$ since $1 - (n-1)\lambda < 1 < 1 + \lambda$.)

If **F** lands on or to the left of F_n, then clearly P decreases by exactly $d > \gamma d$.

On the other hand, suppose **F** lands to the right of F_n. Let A be its starting position, B the point it jumps over, and C its landing position, so that A, B, F_n, C are in that order from left to right. Then P changes by

$$(n-1) F_n C - A F_n,$$

since the distance between each flea (besides **F**) and the rightmost flea increases by $F_n C$ for an total increase of $(n-1) F_n C$; as for **F**, its distance decreases by $A F_n$ since it *becomes* the rightmost flea.

Therefore P changes by at most

$$(n-1) F_n C - A F_n \leq (n-1) BC - AB$$
$$\leq \lambda(n-1) AB - AB = AB[(n-1)\lambda - 1]$$
$$= \frac{AC}{1+\lambda}[(n-1)\lambda - 1] = d\left(\frac{(n-1)\lambda - 1}{1+\lambda}\right),$$

so indeed P decreases by at least

$$d\left(\frac{1 - (n-1)\lambda}{1 + \lambda}\right) = \gamma d$$

in this case as well.

Now suppose we have a configuration O_0 of fleas where the leftmost flea $\mathbf{L}_0$ is at point F_1. Set $P_0 = P(O_0)$, and choose M to the right of F_1 such that

$$F_1 M > \frac{P_0}{\gamma}.$$

Each time $\mathbf{L}_0$ jumps a distance d he decreases P by at least γd, so if it moves a total distance of D he decreases P by at least γD. Because P must always be nonnegative (since it is the sum of nonnegative distances), flea $\mathbf{L}_0$ can decrease P by at most P_0. Thus

$$P_0 \geq \gamma D \text{ and } D \leq \frac{P_0}{\gamma} < F_1 M.$$

It follows that $\mathbf{L}_0$ can never jump to the right of M. Therefore, when $\lambda < 1/(n-1)$, it is *not* always possible to make all the fleas move past M.

Second Solution. (by Ricky Liu) The answer is $\lambda \geq \dfrac{1}{n-1}$.

Let us number the points on the line like those of the real number line, and again assume without loss of generality that the fleas begin at n distinct points. We denote a given position of fleas by the ordered n-tuple $(a_1, a_2, \ldots, a_n)$, where $a_1 < a_2 < \cdots < a_n$ and a_k denotes the position of the kth flea from the left. We define the *value* of a position $(a_1, a_2, \ldots, a_n)$ as

$$V(a_1, a_2, \ldots, a_n, \lambda) = a_n - \lambda(a_1 + a_2 + \cdots + a_{n-1}).$$

We will show that the value of a position of fleas is a monovariant under each move, that is, it does not increase after a move takes place. Assume, for some $j < k$, that the jth flea at a_j jumps over the kth flea at a_k, moving forward a distance $C > 0$ and landing at $a_k + \lambda(a_k - a_j)$.

If he does not land past the rightmost flea, then the value of the position decreases by $\lambda C > 0$.

If he *does* land past the rightmost flea, then the value of the new position is

$$a_k + \lambda(a_k - a_j) - \lambda((a_1 + a_2 + \cdots + a_n) - a_j)$$
$$\leq a_n + \lambda(a_n - a_1) - \lambda(a_2 + a_3 + \cdots + a_n)$$
$$= a_n - \lambda(a_1 + a_2 + \cdots + a_{n-1}) = V(a_1, a_2, \ldots, a_n, \lambda),$$

with equality when $a_k = a_n$. Thus the value of the position never increases, and it stays the same if the leftmost flea jumps over the rightmost flea.

Now, consider an arbitrary position $(a_1, a_2, \ldots, a_n)$. Note that

$$V(a_1, a_2, \ldots, a_n, \lambda) = a_n - \lambda(a_1 + a_2 + \cdots + a_{n-1})$$
$$> a_n - \lambda(a_n + a_n + \cdots + a_n) = a_n - \lambda(n-1)a_n$$
$$= a_n(1 - \lambda(n-1)).$$

Suppose, for the sake of contradiction, that $\lambda < 1/(n-1)$ and that a_n can become arbitrarily large, that is, the nth flea could get past any point M on the line. Because $(1 - \lambda(n-1))$ is positive, $a_n(1 - \lambda(n-1))$ and V must be able to become arbitrarily large. But we know that V is always finite and is nonincreasing. Thus, we have reached a contradiction, so the goal is not possible if $\lambda < 1/(n-1)$.

Now suppose $\lambda \geq 1/(n-1)$, and repeatedly jump the leftmost flea of each position over the rightmost flea. Mark the original locations of the fleas; and each time a flea lands, mark its landing position as well. Then given fixed j, the distance between the jth marked point and the $(j+1)$th marked point is a nondecreasing function of λ; so the *position* of the jth marked point is also a nondecreasing function of λ. Therefore, it suffices to prove that the fleas can pass M for $\lambda = 1/(n-1)$ with this method; because if the j_0th point was to the right of M for $\lambda = 1/(n-1)$, then the j_0th point would also be to the right of M for all $\lambda > 1/(n-1)$.

Now suppose, for the sake of contradiction, that we cannot reach arbitrarily large values for $\lambda = 1/(n-1)$ by repeatedly jumping the leftmost flea of each position over the rightmost flea; remember that in this situation, V remains constant. The location of the current rightmost flea increases after each jump, and by hypothesis, it is bounded; so there must be some limit L, the coordinate of the leftmost unattainable point from a given position. Then the fleas get arbitrarily close to L. Because the value is a continuous function of the a_i, as the fleas approach L, the value of the position must, thus, get closer and closer to the value when all the fleas are located at L, namely $L(1 - \lambda(n-1))$. Since V is constant, this must indeed *be* the value. However, $\lambda = 1/(n-1)$, so the value must be $L(1 - \lambda(n-1)) = 0$.

Since not all the fleas are at a single point, the nth flea is farther to the right than the average of the other $n-1$ fleas. Thus,

$$V(a_1, a_2, \ldots, a_n, \frac{1}{n-1}) = a_n - \frac{1}{n-1}(a_1 + a_2 + \cdots + a_{n-1}) > 0,$$

a contradiction. Hence, our assumption that the fleas would not pass M for $\lambda = 1/(n-1)$ is incorrect and we are done.

4. A magician has one hundred cards numbered 1 to 100. He puts them into three boxes, a red one, a white one and a blue one, so that each box contains at least one card. A member of the audience selects two of the three boxes, chooses one card from each and announces the sum of the numbers on the chosen cards. Given this sum, the magician identifies the

Formal Solutions 45

box from which no card has been chosen. How many ways are there to put all the cards into the boxes so that this trick always works? (Two ways are considered different if at least one card is put into a different box.)

The answer to this problem is 12.

First Solution. Let the color of the number i be the color of the box which contains it. In the sequel, all numbers considered are assumed to be integers between 1 and 100. We consider two cases

1. Assume there is an i such that $i, i+1, i+2$ have three different colors, say **rwb** (red, white, blue). Since $i+(i+3) = (i+1)+(i+2)$, $i+3$ is red. We see that three neighboring different colors determine the color of next number. More over, the pattern repeats: **rwb** is followed by **r**, then **w**, **b**, and so on. The arrangement works backwards, as well: **rwb** is preceded by **b**, and so on.

    So it is enough to assign the colors of 1, 2, 3, and this can be done in six different ways. All these arrangements are indeed good because the pairwise sums of red and white numbers, white and blue numbers, and blue and red numbers give different remainders modulo 3.

2. Assume there are no three neighboring numbers of different colors. Without loss of generality, let number 1 be red. Let i be the smallest non-red number, say white. Let the smallest blue number be k. Since there is no **rwb**, we have $i+1 < k$.

    Suppose that $k < 100$. Since $i+k = (i-1)+(k+1)$, $k+1$ is red. However, in view of $i+(k+1) = (i+1)+k$, $i+1$ has to be blue, which contradicts the fact the smallest blue number is k. Therefore $k = 100$.

    Since $(i-1) + 100 = i + 99$, 99 is white. We now show that 1 is red, 100 is blue, and all the other numbers are white. If $t > 1$ were red, then in view of $t + 99 = (t-1) + 100$, $t-1$ should be blue, but the smallest blue is number 100.

    So the coloring is **rww** $\cdots$ **wwb**, and this is indeed good. If the sum is at most 100, then the missing box is blue; if the sum is 101, then the missing box is white; and if the sum is greater than 101, the missing box is red. The number of such arrangements is again six.

Second Solution. We first claim that 1 and 2 are in different colors. If not, say $1, 2, \ldots, i-1$ are in red, i is white, and j be smallest blue number. We have $i \geq 3$ and $j - 1 \geq i$. But in view of $i + (j-1) = (i-1) + j$,

$j-1$ is white, which leads to the fact that the sum $2+(j-1)=1+j$ does not allow the magician to decide on the unpicked box.

Now let 1 be red, 2 be white, and j be the smallest blue number. We consider the following cases.

1. $j=3$. Since $1+4=2+3$, 4 is red. Similarly, 5 is white, 6 is blue, and so on.

2. $j=100$. Since $2+99=1+100$, 99 is white. If $t>1$ is red, since $t+99=(t-1)+100$, $t-1$ is blue, but 100 is the smallest blue number, a contradiction. So $2,3,\ldots,99$ are all white.

3. $3<j<100$. Since $2+j=1+(j+1)$, $j+1$ is red. Since $3+j=2+(j+1)$, 3 is blue, but j is the smallest blue number, a contradiction.

Therefore there are three choices of colors for 1, two choices for 2, and two choices for 3. Once these choices are made, the colors for the remaining numbers are determined.

Third Solution. (by Ricky Liu) First, note that the condition of the trick being possible is equivalent to saying that no two numbers from different boxes have the same sum as two numbers from a different pair of boxes. We will show that for any $n \geq 4$, there are 12 possible ways to do this, namely permutations of

- $A: \{1,4,7,\ldots\}, \{2,5,8,\ldots\}, \{3,6,9,\ldots\}$;
- $B: \{1\}, \{n\}, \{2,3,4,\ldots,n-1\}$.

First we will show that these arrangements work. In configuration A, the numbers are grouped by their residues modulo 3. Since $0+1 \equiv 1$, $0+2 \equiv 2$, and $1+2 \equiv 0 \pmod 3$, sums from different pairs of boxes will never have the same residue modulo 3, so they can never be the same. In configuration B, notice that the sums from the first and third boxes are $3,4,\ldots,n$, that from the first and second boxes is $n+1$, and those from the second and third boxes are $n+2, n+3, \ldots, 2n-1$, so that there are no repeats. Thus, both these arrangements work. We will now show that these are the only arrangements possible.

The base case is $n=4$. Since there is at least one number in each box, there are only six possibilities, and it is easy to check that the only two that work are the ones given.

Now, assume the statement is true for $n=k$, where $k \geq 4$. Suppose we have a configuration for $k+1$ numbers. There are two cases.

1. The arrangement for $k+1$ numbers consists of $k+1$ in its own box.

   Then, the sums $(k+1)+1, (k+1)+2, \ldots, (k+1)+k$ must not appear between the other two boxes, because they can be attained by a pair that includes $k+1$. Thus, k must be in the same box as 2, or the sum $k+2$ would appear again. Likewise, it must appear in the same box as $3, 4, \ldots, k-1$. These numbers all lie in one box, so the last box must contain the only remaining number, 1. This is configuration B.

2. $k+1$ does not lie in its own box.

   If we remove $k+1$, we must now have a configuration that satisfies the problem statement with k numbers. By the inductive hypothesis, they must be in either configuration A or configuration B.

   Suppose it is A, so the numbers are arranged like

   $$(\{k, k-3, \ldots\}, \{k-1, k-4, \ldots\}, \{k-2, k-5, \ldots\}).$$

   Then $k+1$ cannot go in the first or second group, because $(k+1) + (k-2) = k + (k-1)$. Hence, it must go in the third group, which yields configuration A.

   Suppose, instead, that the k numbers are arranged as in configuration B, say $(\{1\}, \{k\}, \{2, 3, \ldots, k-1\})$. Then $k+1$ cannot go in the first group because $(k+1) + 2 = k + 3$, and it cannot go in the second or third group because $(k+1) + 1 = k + 2$. Thus, with this case the magician's trick is impossible.

Hence, the only possible arrangements are the given A and B, along with permutations. Thus there are $2 \times 6 = 12$ possibilities.

Fourth Solution. We begin with the following lemma.

Lemma. For any set S, let $|S|$ denote the number of the elements in S. Let A and B be sets. Define their sum

$$A + B = \{a + b \mid a \in A, b \in B\}.$$

Then

$$|A + B| \geq |A| + |B| - 1,$$

where equality holds if and only if

1. A and B are arithmetic progressions with equal difference; or
2. $|A|$ or $|B|$ is equal to 1.

Proof. Let $A = \{a_1 < a_2 < \cdots < a_{|A|}\}$ and $B = \{b_1 < b_2 < \cdots < b_{|B|}\}$. The following $|A|+|B|-1$ distinct elements, arranged in increasing order, are in $A + B$:

$$a_1 + b_1 < \cdots < a_1 + b_{|B|} < a_2 + b_{|B|} < \cdots a_{|A|} + b_{|B|}. \quad (*)$$

Therefore $|A+B| \geq |A| + |B| - 1$. Let c_i denote the value of the ith element in list $(*)$.

Assume that $|A + B| = |A| + |B| - 1$ and $|A|, |B| > 1$. For any $1 < i \leq |A|$ and $1 < j \leq |B|$, consider the following list of $|A|+|B|-1$ distinct elements in $A+B$:

$$\underbrace{a_1 + b_1 < \cdots < a_1 + b_{j-1}}_{j-1 \text{ elements}} < \underbrace{a_2 + b_{j-1} < \cdots < a_i + b_{j-1}}_{i-1 \text{ elements}} <$$

$$\underbrace{a_i + b_j < \cdots < a_i + b_{|B|}}_{|B|-j+1 \text{ elements}} < \underbrace{a_{i+1} + b_{|B|} < \cdots < a_{|A|} + b_{|B|}}_{|A|-i \text{ elements}}.$$

This list must be the same as $(*)$. Therefore $a_i + b_{j-1} = c_{i+j-2}$. Likewise $a_{i-1} + b_j = c_{i+j-2}$. Therefore $a_i + b_{j-1} = a_{i-1} + b_j$ and $a_i - a_{i-1} = b_j - b_{j-1}$. Hence both A and B are arithmetic progressions with the same common difference. □

Our problem is equivalent to that of finding a partition of $\{1, 2, \ldots, 100\}$ in three nonempty sets R, W, B such that $R + W$, $W + B$, $B + R$ are pairwise disjoint. By the lemma, this implies that

$$|R+W| + |W+B| + |B+R| \geq 2(|R|+|W|+|B|) - 3 = 197.$$

But $R + W$, $W + B$, $B + R$ must be pairwise disjoint subsets of $\{3, 4, \ldots 199\}$, which has 197 elements, so all equalities hold. In fact, since both $3 = 1 + 2$ and $199 = 99 + 100$ are sums, 1 and 2, 99 and 100 must be in different sets. Since $4 = 1 + 3$ and $198 = 98 + 100$ (note that $4 = 2 + 2$ and $198 = 99 + 99$ are not valid) are sums, 1 and 3, 98 and 100 must be in different sets. Therefore 1 and 100 must each be either in a set of itself or in an arithmetic progression with common difference at least 3. By the lemma we have the following cases.

1. if $|R|, |W|, |B| > 1$, then in the light of the previous argument, each must be arithmetic progressions with common difference $d = 3$. Thus we have **rwbrwb** $\cdots$ and its analogous forms as solutions.

2. if $|R| = 1$ and $|W|, |B| > 1$, then both B and C are arithmetic sequences with the same common difference $d \leq 2$. Thus one of

1 and 100 is in an arithmetic progression with common difference $d = 2$, which is impossible.

3. if $|R| = |B| = 1$ and $|W| = 98$, then $\{A, B\} = \{\{1\}, \{100\}\}$. Therefore we have **rww** $\cdots$ **wb** and its analogous forms as solutions.

5. Determine if there exists a positive integer n such that n has exactly 2000 prime divisors and $2^n + 1$ is divisible by n.

Comment. We will present three solutions here. The first one is very powerful: the lemmas require strong background of number theory and the induction structure in the third lemma is superbly designed. The third solution, the one presented by proposer, was elegant but very tricky. The second solution serves well as an in-between solution.

First Solution. (by Reid Barton) We begin with two lemmas.

Lemma 1. If p^t is an odd prime power and m is an integer relatively prime to both p and $p - 1$, then for any a and b relatively prime to p,
$$a^m \equiv b^m \pmod{p^t} \iff a \equiv b \pmod{p^t}.$$

Proof. Since $(a - b) \mid (a^m - b^m)$, if $p^t \mid (a - b)$ then $p^t \mid (a^m - b^m)$. Conversely, suppose a and b are relatively prime to p and $a^m \equiv b^m \pmod{p^t}$. Since m is relatively prime to both p and $p - 1$, m is relatively prime to $(p-1)p^{t-1} = \phi(p^t)$, so there exists a positive integer k such that $mk \equiv 1 \pmod{\phi(p^t)}$. Then
$$a \equiv a^{mk} = (a^m)^k \equiv (b^m)^k = b^{mk} \equiv b \pmod{p^t}$$
as claimed. $\square$

Lemma 2. If p^t is an odd prime power and a and b are relatively prime to p, then
$$a \equiv b \pmod{p^t} \text{ if and only if } a^p \equiv b^p \pmod{p^{t+1}}.$$

Proof. Let a and b be relatively prime to p, and let $c = a - b$. Let p^e be the highest power of p dividing c. Suppose $e \geq 1$. Observe that
$$a^p - b^p = (c+b)^p - b^p = \sum_{k=0}^{p-1} \binom{p}{k} c^{p-k} b^k.$$
The first term in the sum, c^p, is divisible by p^{pe}, and each of the next $p-2$ terms $\binom{p}{k} c^{p-k} b^k$ ($1 \leq k \leq p-2$) is divisible by p^{2e+1} since p divides

$\binom{p}{k}$ and p^{2e} divides c^{p-k}. Thus the sum of all the terms but the last is divisible by p^{e+2} (since $pe \geq e+2$ and $2e+1 \geq e+2$). The last term is divisible by p^{e+1} but not p^{e+2}, since c is divisible by p^e but not p^{e+1} and b is relatively prime to p. So the entire sum $a^p - b^p$ is divisible by p^{e+1} but not p^{e+2}.

Now suppose $a \equiv b \pmod{p^t}$. Then $p^t \mid c = a - b$, so $e \geq t$. Thus, because $a^p - b^p$ is divisible by p^{e+1}, it is divisible by p^{t+1}. Conversely, suppose $a^p \equiv b^p \pmod{p^{t+1}}$. By Fermat's Little Theorem $a \equiv a^p \equiv b^p \equiv b \pmod{p}$, so $p \mid c = a - b$ and $e \geq 1$. Now since p^{t+1} divides $a^p - b^p$ but p^{e+2} does not, we must have $t + 1 < e + 2$ so $e \geq t$. Thus p^t divides $a - b$, as desired. □

We now solve the problem with a strengthened induction statement.

Lemma 3. For each $k \geq 1$, there exists an integer $n_k = p_1^{e_1} \cdots p_k^{e_k} \geq 4$ ($p_1 < \cdots < p_k$ are odd primes and e_i are positive integers) such that

(i) $n_k \mid (2^{n_k} + 1)$;

(ii) for $1 \leq i \leq k$, $p_i^{2e_i+1} \nmid (2^{n_k} + 1)$;

(iii) if q is an odd prime less than p_k which does not divide n_k, then
$q \nmid (2^{n_k} + 1)$.

Proof. For $k = 1$, we can take $n_1 = 3^2 = 9$ since $9 \mid 27 \cdot 19 = 513 = 2^9 + 1$, $3^5 = 243 \nmid 513$, and (iii) holds trivially. Suppose we have an integer n_k satisfying conditions (i), (ii), (iii) for $k \geq 1$; we construct a new integer n_{k+1} such that (i), (ii), (iii) hold with k replaced by $k + 1$. Call these new statements (i'), (ii'), (iii').

Consider the factorization of $2^{n_k} + 1$. By (iii), $p_1, \ldots, p_k$ are the k smallest primes dividing $2^{n_k} + 1$, so we can write $2^{n_k} + 1 = p_1^{c_1} \cdots p_k^{c_k} q_1^{d_1} \cdots q_r^{d_r}$ with $p_1 < \cdots < p_k < q_1 < \cdots < q_r$ for some $c_i \geq 1$, $d_j \geq 1$, $r \geq 0$. By (ii), $c_i \leq 2e_i$ for each i, so $p_1^{c_1} \cdots p_k^{c_k} \leq n_k^2$. Since $2^{n_k} + 1 > n_k^2$ for $n_k > 4$, we must have $r \geq 1$. Define $p = p_{k+1} = q_1$, $e = e_{k+1} = d_1$, $n_{k+1} = n_k p^e = p_1^{e_1} \cdots p_{k+1}^{e_{k+1}}$. By (i), $c_i \geq e_i$ for each i, so $2^{n_k} + 1$ is divisible by n_{k+1}. Thus $2^{n_k} \equiv -1 \pmod{n_{k+1}}$. Raising both sides to the p^e gives $2^{n_{k+1}} \equiv 2^{n_k p^e} \equiv (-1)^{p^e} \equiv -1 \pmod{n_{k+1}}$, so n_{k+1} divides $2^{n_{k+1}} + 1$ and (i') holds.

We now prove (ii'). Suppose $1 \leq i \leq k$ and $p_i^{2e_i+1} \mid (2^{n_{k+1}} + 1)$; then $2^{n_k p^e} \equiv 2^{n_{k+1}} \equiv -1 \equiv (-1)^{p^e} \pmod{p_i^{2e_i+1}}$. Since $p = q_1 > p_k \geq p_i$ is prime, p^e is relatively prime to p_i and $p_i - 1$, so by lemma 1, $2^{n_k} \equiv -1 \pmod{p_i^{2e_i+1}}$. But then $p_i^{2e_i+1} \mid 2^{n_k}$, contradicting (ii). Thus (ii') holds for $1 \leq i \leq k$. It remains to show that $p^{2e+1} \nmid (2^{n_{k+1}} + 1)$; suppose

$2^{n_{k+1}} \equiv -1 \pmod{p^{2e+1}}$. Then by lemma 2
$$2^{n_k p^e} \equiv (-1)^{p^e} \pmod{p^{2e+1}},$$
$$2^{n_k p^{e-1}} \equiv (-1)^{p^{e-1}} \pmod{p^{2e}},$$
...
$$2^{n_k} \equiv -1 \pmod{p^{e+1}},$$
a contradiction, as p^e is the largest power of p dividing $2^{n_k}+1$. Thus (ii') holds for $i = k+1$ as well.

Finally we prove (iii'). Suppose $q < p_{k+1} = p$ is an odd prime dividing $(2^{n_{k+1}} + 1)$ but not n_{k+1}. Then $2^{n_k p^e} \equiv 2^{n_{k+1}} \equiv (-1)^{p^e} \pmod{q}$. Since $p > q$ is a prime, p is relatively prime to both q and $q-1$, so by lemma 1, $2^{n_k} \equiv -1 \pmod{q}$. Therefore $q \mid 2^{n_k} + 1$, so q is one of the primes $p_1, \ldots, p_k, q_1, \ldots, q_r$. But the primes $p_1, \ldots, p_k$ all divide n_{k+1}, so $q \geq q_1 = p$, a contradiction. Thus (iii') holds. By induction lemma 3 holds for all k. □

Now taking $k = 2000$, lemma 3 gives us an integer n_{2000} which is divisible by exactly 2000 distinct primes $p_1, \ldots, p_{2000}$ and such that n_{2000} divides $2^{n_{2000}} + 1$.

Second Solution. (by Zhongtao Wu, China) We begin with three simple lemmas.

Lemma 4. The only positive integer solution for the equation $2^x + 1 = 3^y$ are $(x, y) = (1,1), (3,2)$.

Proof. Suppose that (x, y) is a solution. If $x = 1$ then $y = 1$; if $x > 1$, then
$$1 \equiv 2^x + 1 \equiv 3^y \equiv (-1)^y \pmod{4}.$$
Thus y is even. Let $y = 2y_1$. Then
$$2^x = 3^y - 1 = 3^{2y_1} - 1 = (3^{y_1} + 1)(3^{y_1} - 1).$$
So $3^{y_1} + 1 = 2^{x_1}$ and $3^{y_1} - 1 = 2^{x_2}$. Since $2^{x_1} - 2^{x_2} = 2$, $x_1 = 2$, and $x_2 = 1$. Therefore $y_1 = 1$, $x = x_1 + x_2 = 3$, and $y = 2y_1 = 2$. □

Lemma 5. Let n, a, b be positive integers. Then
$$\gcd(n^a - 1, n^b - 1) = n^{\gcd(a,b)} - 1.$$

Proof. Since $\gcd(a, b)$ divides both a and b,
$$(n^{\gcd(a,b)} - 1) \mid (n^a - 1) \text{ and } (n^{\gcd(a,b)} - 1) \mid (n^b - 1).$$

Thus $n^{\gcd(a,b)} - 1 \mid \gcd(n^a - 1, n^b - 1)$.

On the other hand, there exist positive integers x and y such that $ax - by = \gcd(a,b)$. Then $(n^a - 1) \mid (n^{ax} - 1)$ and $(n^b - 1) \mid (n^{by} - 1)$. Note that

$$n^{ax} - 1 = n^{by}(n^{\gcd(a,b)} - 1) + n^{by} - 1.$$

Since $\gcd(n^{by}, n^b - 1) = 1$, $\gcd(n^a - 1, n^b - 1) \mid (n^{\gcd(a,b)} - 1)$. Therefore $\gcd(n^a - 1, n^b - 1) = n^{\gcd(a,b)} - 1$.

Lemma 6. Let c, a, b be positive integers such that $c \mid (2^a + 1)$ and $c \mid (2^b + 1)$. Then $c \mid (2^{\gcd(a,b)} + 1)$.

Proof. Note that

$$\gcd(2^a + 1, 2^a - 1) = \gcd(2^b + 1, 2^b - 1) = 1,$$

Setting $n = 2$ in lemma 5 gives

$$(2^{\gcd(a,b)} - 1)(2^{\gcd(a,b)} + 1) = 2^{2\gcd(a,b)} - 1$$
$$= 2^{\gcd(2a,2b)} - 1 = \gcd(2^{2a} - 1, 2^{2b} - 1)$$
$$= \gcd[(2^a - 1)(2^a + 1), (2^b - 1)(2^b + 1)]$$
$$= \gcd(2^a - 1, 2^b - 1) \cdot \gcd(2^a - 1, 2^b + 1)$$
$$\cdot \gcd(2^a + 1, 2^b - 1) \cdot \gcd(2^a + 1, 2^b + 1)$$
$$= (2^{\gcd(a,b)} - 1)\gcd(2^a - 1, 2^b + 1)$$
$$\cdot \gcd(2^a + 1, 2^b - 1) \cdot \gcd(2^a + 1, 2^b + 1).$$

Therefore $\gcd(2^a + 1, 2^b + 1) \mid (2^{\gcd(a,b)} + 1)$. This completes the proof of lemma 6. □.

We construct inductively an infinite sequence of distinct primes $\{p_k\}$ so that $p_1 = 3$, $p_2 = 19$, and for $k \geq 3$, we have $p_k \mid (2^{p_{k-1}} + 1)$.

Suppose that odd distinct primes $p_1 = 3, p_2 = 19, \ldots, p_k$, $k \geq 2$, are defined. By lemma 4 there is an prime divisor $p \neq 3$ of $2^{p_k} + 1$. We consider the following cases.

1. if $p = p_k$, then $p_k \mid (2^{p_k} + 1)$. By **Fermat's little theorem**, $p_k \mid (2^{p_k - 1} - 1)$. Therefore

$$p_k \mid [(2^{p_k} + 1) - 2(2^{p_k - 1} - 1)],$$

which implies that $p_k = 3$, a contradiction.

2. if $p = p_i$ for some $3 \leq i \leq k-1$, then $p_i \mid (2^{p_k}+1)$. By the definition of p_i, $p_i \mid (2^{p_{i-1}}+1)$. By lemma 6, $p_i \mid (2^{\gcd(p_k,p_{i-1})}+1) = 3$, a contradiction.

3. if $p = p_2 = 19$, then $19 \mid (2^{p_k}+1)$. Since $19 \mid (2^9+1)$, by lemma 6, $p \mid (2^{\gcd(p_k,9)}+1) = 3$, a contradiction.

Therefore $p \neq p_1, p_2, p_3, \ldots, p_k$. Now we let $p_{k+1} = p$ and this completes our induction.

Let $n = 3^2 \cdot p_2 \ldots p_{2000}$. Then n has 2000 prime divisors. For $3 \leq i \leq 2000$, $p_i \mid 2^{p_{i-1}}+1$. Since n is odd and $p_i \mid n$, $(2^{p_i}+1) \mid (2^n+1)$. Then $p_i \mid (2^n+1)$. Note also that both 3^2 and $p_2 = 19$ divide $2^9 + 1$, which is a divisor of $2^n + 1$. Therefore $n \mid (2^n + 1)$, as desired.

Third Solution. We start with the following lemma.

Lemma 7. For any integer $a > 2$ there exists a prime p such that $p \mid (a^3 + 1)$ but $p \nmid (a+1)$.

Proof. For the sake of contradiction assume the statement is false for some integer $a > 2$. Since $a^3 + 1 = (a+1)(a^2 - a + 1)$, each prime divisor of $a^2 - a + 1$ divides $a + 1$. The identity

$$a^2 - a + 1 = (a+1)(a-2) + 3 \qquad (1)$$

then shows that 3 is the only prime dividing $a^2 - a + 1$, that is, $a^2 - a + 1$ is a power of 3. Since $3 \mid (a+1)$, $3 \mid (a-2)$. Hence the right-hand side of (1) is divisible by 3 but not by 9. Being a power of 3, $a^2 - a + 1 = 3$ and $a = 2$, a contradiction. Therefore our assumption is false and the original statement is true. □

We now prove by induction on k a more general statement:

For each $k \in \mathbb{N}$ there exists $n = n_k \in \mathbb{N}$ such that $n \mid (2^n + 1)$, $3 \mid n$, and n has exactly k prime divisors.

The base case of the induction can be $n_1 = 3$.

For the inductive step, assume that for certain $k \geq 1$ there is a number $n_k = 3^\ell t$, $\ell \geq 1$ and $3 \nmid t$, satisfying the conditions. Then n_k is odd, and hence $3 \mid 2^{2n_k} - 2^{n_k} + 1$. The equality

$$2^{3n_k} + 1 = (2^{n_k} + 1)(2^{2n_k} - 2^{n_k} + 1)$$

shows that $3^{\ell+1} \mid (2^{3n_k} + 1)$. Lemma 7 shows that there is an odd prime p such that $p \mid (2^{3n_k} + 1)$ and $p \nmid (2^{n_k} + 1)$. Then $n_{k+1} = 3n_k p$ satisfies the requirements for $k+1$. This completes the induction.

Note. Indeed, in light of lemma 7, one can just set $n = 3^{2000} \cdot p_2 \cdots p_{2000}$ where $p_2 \neq 3$, $p_2 \mid (2^{3^2} + 1)$, $p_{i+1} \mid (2^{3^{i+1}} + 1)$ and $p_{i+1} \nmid (2^{3^i} + 1)$, for $i = 2, \ldots, 2000$.

6. Let $\overline{AH_1}$, $\overline{BH_2}$, and $\overline{CH_3}$ be the altitudes of an acute triangle ABC. The incircle ω of triangle ABC touches the sides BC, CA, and AB at T_1, T_2, and T_3, respectively. Consider the symmetric images of the lines H_1H_2, H_2H_3, and H_3H_1 with respect to the lines T_1T_2, T_2T_3, and T_3T_1. Prove that these images form a triangle whose vertices lie on ω.

Comment. This is another very difficult problem. During the Jury meetings before the IMO, one leader commented that "this problem should be submitted to a journal" (to allow people several weeks to work on it). Before the exam, there was a poll among leaders/leader observers predicting the number of complete solutions on this problem. Most people expected about 10–12 complete solutions. It turned out that there were 33 complete solutions, which exceeded the most optimistic expectations. We present four solutions. The first two solutions use symmetry across the angle bisector; the other two use **homothety**. Skillful calculations are an important aspect of problem solving. Indeed, almost all solutions given by the students at the IMO involved some degree of calculations.

First Solution. (by Zhiwei Yun, China) Since the obtained triangle is unique, we can work backward. The desired result follows from the lemma and its analogous statements.

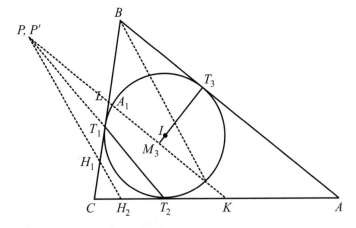

Lemma 1. Let $A_1B_1C_1$ be an acute triangle inscribed in circle ω. Line AB is parallel to $\overline{A_1B_1}$ and is tangent to ω. (Among two such possible

lines, the line AB is further from $\overline{A_1B_1}$.) Likewise define the lines BC and CA. Let $\overline{AH_1}$, $\overline{BH_2}$, and $\overline{CH_3}$ be the altitudes of triangle ABC. The incircle ω of triangle ABC touches the sides BC, CA, and AB at T_1, T_2, and T_3, respectively. Then the lines A_1B_1 and H_1H_2 are symmetric across the line T_1T_2.

Proof. Let A, B, C, a, b, c, s, I, r denote the angles, side lengths, semiperimeter, incenter, inradius of triangle ABC. In the sequel, we assume $A \leq B \leq C$; other configurations only require slight modifications. Let the line A_1B_1 meet $\overline{CA}$ and $\overline{BC}$ at K and L, respectively. Since $A + B + C = 180°$ and $\overline{A_1B_1} \parallel \overline{AB}$,

$$\angle CH_2H_1 - \angle CT_2T_1 = B - (90° - C/2)$$
$$= (90° - C/2) - A = \angle CT_2T_1 - \angle CKA_1.$$

Therefore the lines H_1H_2 and A_1B_1 are on the different sides of the line T_1T_2 and form the same angle with the line T_1T_2. So we only need to prove that the lines T_1T_2, H_1H_2, and A_1B_1 are concurrent, i.e., $P = P'$, where P and P' are the respective intersections of the line T_1T_2 with the lines H_1H_2 and A_1B_1.

Let M_3 be the midpoint of $\overline{A_1B_1}$. Then T_3, I, M_3 are collinear. Since $\angle B_1IM_3 = \angle B_1IA_1/2 = \angle B_1C_1A_1 = C$, we have $T_3M_3 = T_3I + IM_3 = r + r\cos C$. Since triangles CKL and CAB are similar, the similitude is also the ratio between corresponding altitudes

$$\frac{CK}{CA} = \frac{CL}{CB} = \frac{CH_3 - T_3M_3}{CH_3} = \frac{\frac{2rs}{c} - r(1 + \cos C)}{\frac{2rs}{c}}$$

$$= 1 - \frac{c(1 + \cos C)}{2s} = 1 - \frac{c\left(1 + \frac{a^2+b^2-c^2}{2ab}\right)}{2s}$$

$$= 1 - \frac{c[(a+b)^2 - c^2]}{2ab(a+b+c)} = 1 - \frac{c(a+b-c)}{2ab} = \lambda.$$

Hence $CK = b\lambda$ and $CL = a\lambda$. Applying the **Menelaus' theorem** to triangle CT_1T_2 and the line H_1H_2P, triangle CT_1T_2 and the line KLP' (line A_1B_1) gives

$$\frac{CH_2}{H_2T_2} \cdot \frac{T_2P}{PT_1} \cdot \frac{T_1H_1}{H_1C} = 1 \iff \frac{T_2P}{PT_1} = \frac{H_1C \cdot T_2H_2}{H_1T_1 \cdot H_2C},$$

$$\frac{CK}{KT_2} \cdot \frac{T_2P'}{P'T_1} \cdot \frac{T_1L}{LC} = 1 \iff \frac{T_2P'}{P'T_1} = \frac{LC \cdot KT_2}{CK \cdot T_1L}.$$

Note that to prove $P = P'$, it is enough to show that $T_2P/PT_1 = T_2P'/P'T_1$. So we only need to prove that

$$\frac{H_1C \cdot T_2H_2}{H_1T_1 \cdot H_2C} = \frac{LC \cdot KT_2}{CK \cdot T_1L}.$$

This equivalent to

$$\frac{b\cos C[a\cos C - (s-c)]}{a\cos C[b\cos C - (s-c)]} = \frac{a\lambda(s-c-b\lambda)}{b\lambda(s-c-a\lambda)},$$

$$\frac{a^2}{b^2} = \frac{[a\cos C - (s-c)]}{[b\cos C - (s-c)]} \cdot \frac{s-c-a\lambda}{s-c-b\lambda},$$

which follows from

$$a\cos C - (s-c) = \frac{a^2+b^2-c^2}{2b} - \frac{a+b-c}{2}$$

$$= \frac{a^2-c^2-b(a-c)}{2b} = \frac{(a-c)(a+c-b)}{2b}$$

and

$$s - c - a\lambda = \frac{a+b-c}{2} - a + \frac{c(a+b-c)}{2b}$$

$$= \frac{ab+b^2-cb-2ab+ca+cb-c^2}{2b}$$

$$= \frac{b^2-c^2+ca-ab}{2b} = \frac{(b-c)(b+c-a)}{2b},$$

and the analogous equalities

$$b\cos C - (s-c) = \frac{(b-c)(b+c-a)}{2a},$$

$$s - c - b\lambda = \frac{(a-c)(a+c-b)}{2a}.$$

This completes the proof of the lemma. □

Second Solution.

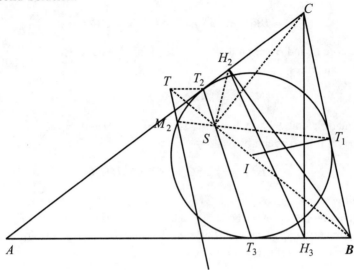

Let A_1, B_1, C_1 be the reflections of T_1, T_2, T_3 across the bisectors of $\angle A$, $\angle B$, $\angle C$, respectively. Then A_1, B_1, C_1 lie on ω. We prove that they are the vertices of the triangle formed by the images in question, which settles the claim.

By symmetry, it suffices to show that the reflection ℓ_1 of the line $H_2 H_3$ across the line $T_2 T_3$ passes through B_1. Let I be the center of ω. Note that T_2 and H_2 are always on the same side of the line BI, with T_2 closer to the line BI than that of H_2. In the sequel, we consider only the case when C is on the same side of the line BI, as in the figure, i.e., $\angle C \geq \angle A$ (minor modifications are needed if C is on the other side).

Let $\angle A = 2\alpha$, $\angle B = 2\beta$, $\angle C = 2\gamma$. Then $\alpha + \beta + \gamma = 90°$.

Lemma 2. The mirror image of H_2 with respect the line $T_2 T_3$ lies on the line BI.

Proof. Let ℓ be the line passing through H_2 and perpendicular to the line $T_2 T_3$. Denote by S and T the points of intersections of the line BI with $\overline{T_2 T_3}$ and ℓ. Note that S also lies on $\overline{BT}$. It is sufficient to prove that $\angle TSH_2 = 2\angle TST_2$.

We have

$$\angle TST_2 = \angle BST_3 = \angle AT_3 S - \angle T_3 BS = (90° - \alpha) - \beta = \gamma.$$

By symmetry across the line BI, $\angle BST_1 = \angle BST_3 = \gamma$. Note that

$$\angle BT_1 S = 180° - \angle BST_1 - \angle T_1 BS = 90° + \alpha > 90°.$$

Therefore C and S are on the same side of the line IT_1. Then, in the view of the equalities

$$\angle IST_1 = \angle BST_1 = \gamma = \angle ICT_1,$$

the quadrilateral SIT_1C is cyclic, so $\angle ISC = \angle IT_1C = 90°$. But then BCH_2S is also cyclic by

$$\angle BSC = \angle ISC = 90° = \angle BH_2C.$$

It follows that $\angle TSH_2 = \angle BCH_2 = 2\gamma = 2\angle TST_2$, as desired. □

Note that the proof of the lemma 2 also gives $\angle BTT_2 = \angle SH_2T_2 = \beta$, by symmetry across the line T_2T_3 and because BCH_2S is cyclic. Then, since B_1 is the reflection of T_2 across the line BI,

$$\angle BTB_1 = \angle BTT_2 = \beta = \angle CBT$$

and $TB_1 \parallel BC$. To prove that B_1 lies on ℓ_1, it now suffices to show that $\ell_1 \parallel BC$.

Suppose that $\beta \neq \gamma$ (otherwise it is trivial that $\ell_1 \parallel BC$); let the line CB meet the lines H_2H_3 and T_2T_3 at D and E, respectively. Note that D and E lie on the line BC on the same side of $\overline{BC}$. We have $\angle BDH_3 = 2|\beta - \gamma|$ and $\angle BET_3 = |\beta - \gamma|$. Therefore $\ell_1 \parallel BC$. The proof is complete.

Third Solution. Let $\angle A = 2\alpha$, $\angle B = 2\beta$, $\angle C = 2\gamma$. It is easy to compute that angles between the line AB and the lines H_1H_2 and T_1T_2 are $2|\alpha - \beta|$ and $|\alpha - \beta|$, respectively. It follows that the mirror image of the line H_1H_2 across the line T_1T_2 is parallel to the line AB. A similar conclusion holds for the other mirror images. So the triangle $A_1B_1C_1$ formed by the three reflection lines has sides parallel to those of triangle ABC. Hence there is a **homothety H** taking triangle ABC to $A_1B_1C_1$. Now the claim is that $A_1B_1C_1$ is inscribed in ω. This can be true if and only if **H** takes the circumcircle of triangle ABC to ω.

Let ω', O, I, R, r be the circumcircle, circumcenter, incenter, circumradius, inradius of triangle ABC. Let J be the point dividing $\overline{OI}$ internally in ratio $OJ : JI = R : r$. Consider the homothety with center J and magnitude $-r/R$. This homothety takes ω' to ω, and so coincides with

Formal Solutions

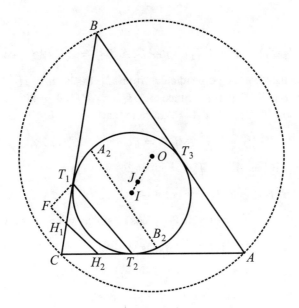

H. Let $\mathbf{H}(ABC) = A_2B_2C_2$. It suffices to show that the line A_2B_2 is the reflection of the lines H_1H_2 across the lines T_1T_2.

Let the directed distance $d(\mathcal{A}, \mathcal{B})$ between two objects $\mathcal{A}$ and $\mathcal{B}$ (either a line and a point, or two parallel lines) be the directed length of the shortest segment connecting these two objects. (By *directed lengths of segments*, we mean that if rays PQ and RS point in the same direction, then PQ and RS have the same sign while PQ and SR have opposite signs.) Then the points T that satisfy $d(T, H_1H_2) = -d(T, A_2B_2)$ are all equidistant from lines H_1H_2 and A_2B_2, and in fact lie on the *same* bisector of the angle formed by these two lines. (If the two lines are parallel, the points T lie on the line midway between them). Thus it suffices to prove that $d(T_1, H_1H_2) = -d(T_1, A_2B_2)$ and $d(T_2, H_1H_2) = -d(T_2, A_2B_2)$.

In this solution, we choose positive directions so that $d(AB, C)$, $d(A, C)$, $d(B, C)$, and $D(H_1H_2, C)$ are all positive. Then

$$d(I, A_2B_2) = \frac{r}{R} \cdot d(AB, O) = \frac{r}{R} \cdot R\cos 2\gamma = r\cos 2\gamma$$

and $d(AB, A_2B_2) = r(1 + \cos 2\gamma)$. Also,

$$d(AB, T_1) = BT_1 \sin 2\beta = r \cot \beta (2\sin\beta\cos\beta)$$
$$= 2r\cos^2\beta = r(1 + \cos 2\beta).$$

Hence
$$d(A_2B_2, T_1) = d(AB, T_1) - d(AB, A_2B_2) = r(\cos 2\beta - \cos 2\gamma).$$

Let F be the foot of perpendicular from T_1 to the line H_1H_2. Then $T_1F = T_1H_1 \sin \angle T_1H_1F$. Since $\angle AH_1B = BH_2A = 90°$, ABH_1H_2 is cyclic, so $\angle T_1H_1F = \angle A = 2\alpha$. Note that
$$T_1H_1 = T_1C - H_1C = r\cot\gamma - AC\cos 2\gamma.$$

Expressing AC as $AT_2 + T_2C = r(\cot 2\alpha + \cot\gamma)$ gives
$$\begin{aligned}
d(T_1, H_1H_2) = T_1F &= (r\cot\gamma - AC\cos 2\gamma)\sin 2\alpha \\
&= r[\cot\gamma \sin 2\alpha - \cos 2\gamma \sin 2\alpha(\cot\gamma + \cot\alpha)] \\
&= r[\sin 2\alpha \cot\gamma(1 - \cos 2\gamma) - \cos 2\gamma \sin 2\alpha \cot\alpha] \\
&= r[\sin 2\alpha \cot\gamma(2\sin^2\gamma) - 2\cos^2\alpha \cos 2\gamma] \\
&= r[\sin 2\alpha \sin 2\gamma - (1 + \cos 2\alpha)\cos 2\gamma] \\
&= r[\sin 2\alpha \sin 2\gamma - \cos 2\alpha \cos 2\gamma - \cos 2\gamma] \\
&= r[-\cos(2\alpha + 2\gamma) - \cos 2\gamma] \\
&= r(\cos 2\beta - \cos 2\gamma) = -d(T_1, A_2B_2),
\end{aligned}$$
as desired. Likewise, $d(T_2, H_1H_2) = r(\cos 2\alpha - \cos 2\gamma) = -d(T_2, A_2B_2)$. The proof is complete.

Fourth Solution. (by Kiran Kedlaya) Let H and I be the orthocenter and incenter, respectively, of triangle ABC. Since $\angle BH_2C = \angle CH_3B = 90°$, BH_3H_2C is cyclic, so $\angle AH_2H_3 = \angle ABC$. Therefore triangles AH_2H_3 and ABC are oppositely similar. In particular, reflecting the line H_2H_3 across the line T_2T_3, which is perpendicular to the angle bisector AI of A, gives a line parallel to BC.

Therefore the triangle formed by the reflections has sides parallel to the sides of ABC. By looking at the desired result, we realize that it suffices to show that these reflections form the triangle obtained from ABC by the **homothety** with negative ratio taking the circumcircle of ABC to its incircle. (We take the ratio to be negative because that gives the correct assertion in case ABC is equilateral.) In particular, it suffices to show that the reflection of the line H_2H_3 across the line T_2T_3 intersects ω obtaining a chord C_1B_1 parallel to the line BC, between A and the incenter I, which intercepts an arc of measure $2\angle A$.

Formal Solutions

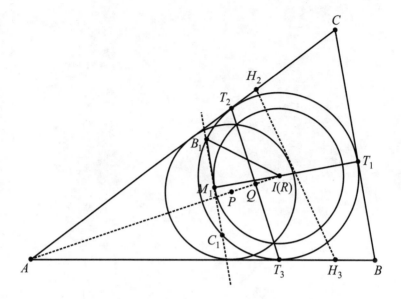

The coefficient of similitude between AH_2H_3 and ABC is $AH_3/AC = \cos A$. That means we can obtain AH_2H_3 from ABC by **dilating** towards A with ratio $\cos A$, then reflecting across AI. In particular, the line H_2H_3 is tangent to the circle ω_1, the incircle of triangle AH_2H_3, obtained from the incircle of triangle ABC by **dilating** towards A with ratio of $\cos A$. Let P be the center of ω_1, Q the intersection of the line AI with the line T_2T_3, and R the reflection of P across the line T_2T_3. Then

$$AP = AI\cos A,$$
$$AQ = AT_3 \cos A/2 = AI \cos^2 A/2,$$
$$AR = 2AQ - AI = AI(2\cos^2 A/2 - \cos A) = AI.$$

Let **T** denote the reflection across the line T_2T_3. Under **T**, the respective images of the line H_2H_3, ω_1 (radius $r\cos A$), and its center P are the line C_1B_1, circle ω_2 (radius $r\cos A$), and its center I. Since the line H_2H_3 is tangent to ω_1, the line C_1B_1 is tangent to ω_2, i.e., the distance from chord C_1B_1 to I is $r\cos A$. Therefore $C_1B_1 = 2r\sin A$, and so intercepts an arc of measure $2\angle A$ by the **Extended Law of Sines**. Moreover, the line H_2H_3 and A lie on opposite sides of P, so BC and B_1C_1 lie on opposite sides of O.

Thus as noted above, the reflection of the line H_2H_3 contains the image of $\overline{BC}$ under the homothety of negative ratio taking the circumcircle of ABC to its incircle, which suffices to prove the desired result.

4 Problem Credits

USAMO

1. Bjorn Poonen
2. Titu Andreescu
3. Richard Stong
4. Alexander Soifer
5. Kiran Kedlaya
6. Gheorghita Zbaganu (Romania)

Team Selection Test

1. Titu Andreescu
2. Titu Andreescu
3. Kiran Kedlaya
4. from *Kvant*
5. Kiran Kedlaya
6. Titu Andreescu

IMO

1. Russia
2. USA
3. Russia
4. Hungary
5. Russia
6. Belarus

5
Glossary

AM-GM Inequality If $a_1, a_2, \ldots, a_n$ are n nonnegative numbers, then
$$\frac{1}{n}\sum_{i=1}^{n} a_i \geq (a_1 a_2 \cdots a_n)^{\frac{1}{n}},$$
with equality if and only if $a_1 = a_2 = \cdots = a_n$.

Binomial Coefficient
$$\binom{n}{k} = \frac{n!}{k!(n-k)!},$$
the coefficient of x^k in the expansion of $(x+1)^n$.

Dilation See **homothety**.

Extended Law of Sines In a triangle ABC with circumradius equal to R,
$$\frac{\sin A}{BC} = \frac{\sin B}{AC} = \frac{\sin C}{AB} = 2R.$$

Erdős-Mordell Inequality Let P be a point inside of triangle ABC, and let Q, R, and S be the feet of perpendiculars from P to BC, CA, and AB, respectively. Then
$$PA + PB + PC \geq 2(PQ + PR + PS),$$
with equality if and only if triangle ABC is equilateral and P is its center.

Fermat's Little Theorem If p is prime, then $a^p \equiv a \pmod{p}$, for all integers a.

Heron's Formula The area of a triangle ABC with sides a, b, c is equal to
$$[ABC] = \sqrt{s(s-a)(s-b)(s-c)},$$
where $s = (a+b+c)/2$.

Further suppose that P, Q, R are the points of tangency of the incircle with sides AB, BC, CA, respectively. Let $AP = x$, $BQ = y$, $CR = z$. Then $AR = x$, $BP = y$, $CQ = z$, and
$$x = s-a, \ y = s-b, \ y = s-c, \ x+y+z = s.$$
Therefore Heron's formula reads
$$[ABC] = \sqrt{xyz(x+y+z)}.$$

Homothety A homothety (**dilation, central similarity**, also see **transformation**) is a transformation that fixes one point O (its center) and maps each point P to a point P' for which O, P, P' are collinear and the ratio $OP : OP' = k$ is constant (k can be either positive or negative), where k is called the **magnitude** (**similitude, ratio**) of the homothety.

Inversion of Center O and Ratio r Given a point O in the plane and a real number $r > 0$, the inversion through O with radius r maps every point $P \neq O$ to the point P' on the ray $\overrightarrow{OP}$ such that $OP \cdot OP' = r^2$. We also refer to this map as inversion through ω, the circle with center O and radius r. Key properties of inversion are:

1. Lines through O invert to themselves (though the individual points on the line are not all fixed).
2. Lines not through O invert to circles through O and vice versa.
3. Circles not through O invert to other circles not through O.
4. A circle other than ω inverts to itself (as a whole, not point-by-point) if and only if it is orthogonal to ω, that is, it intersects ω and the tangents to the circle and to ω at either intersection point are perpendicular.

Kite A quadrilateral with its sides forming two pairs of congruent adjacent sides. A kite is symmetric with one of its diagonals. (If it is symmetric with both diagonals, it becomes a rhombus.) Two diagonals of a kite are perpendicular to each other. For example, if $ABCD$ is a quadrilateral with $AB = AD$ and $CB = CD$, then $ABCD$ is a kite and is the symmetric with the diagonal AC.

Glossary

Menelaus' Theorem Let ABC be a triangle, and let P, Q, R be points on the lines BC, CA, AB, respectively. Then P, Q, R are collinear if and only if
$$\frac{BP}{PC}\frac{CQ}{QA}\frac{AR}{RB} = 1.$$
(If the lengths are directed, then the product is -1.)

Power of a Point Theorem Given a fixed point P and a fixed circle ω, draw a line through P which intersects the circle at X and Y. The power of the point P with respect to ω is defined to be $PX \cdot PY$. The power of a point theorem states that this quantity is a constant; i.e., does not depend on which line was drawn through P. Note that it does not matter whether P was in, on or outside ω.

Schur's Inequality Let x, y, z be nonnegative real numbers. Then for any $r > 0$,
$$x^r(x-y)(x-z) + y^r(y-z)(y-x) + z^r(z-x)(z-y) \geq 0.$$
Equality holds if and only if $x = y = z$ or if two of x, y, z are equal and the third is equal to 0.

The proof of the inequality is rather simple. Since the inequality is symmetric in the three variables, we may assume without loss of generality that $x \geq y \geq z$. Then the given inequality may be rewritten as
$$(x-y)[x^r(x-z) - y^r(y-z)] + z^r(x-z)(y-z) \geq 0,$$
and every term on the left-hand side is clearly nonnegative. The first term is positive if $x > y$, so equality requires $x = y$, as well as $z^r(x-z)(y-z) = 0$, which gives either $x = y = z$ or $z = 0$.

Spiral Similarity See **transformation**.

Transformation A mapping of the plane onto itself such that every point P is mapped into a unique image P' and every point Q' has a unique prototype (preimage, inverse image, counterimage) Q.

A **reflection across a line** (in the plane) is a transformation which takes every point in the plane into its mirror image, with the line as mirror. A **rotation** is a transformation resulting from the entire plane being rotated about a fixed point in the plane.

A **similarity** is a transformation that preserves ratios of distances. If P' and Q' are the respective images of points P and Q under a similarity $\mathbf{T}$, then the ratio $P'Q'/PQ$ depends only on **T**. This ratio is the **similitude** of **T**. A **dilation** is a direction-preserving similarity, i.e., a similarity that takes each line into a parallel line.

The **product of two transformations** is the result of applying the first transformation and then the second. A **spiral similarity** is the product of a rotation and a dilation, or vice versa.

Trigonometric Identities

$$\sin^2 x + \cos^2 x = 1,$$

$$\tan x = \frac{\sin x}{\cos x}, \quad \cot x = \frac{1}{\tan x},$$

$$\sin(-x) = -\sin x, \quad \cos(-x) = \cos(x),$$

$$\tan(-x) = -\tan x, \quad \cot(-x) = -\cot x,$$

$$\sin(90° \pm x) = \cos x, \quad \cos(90° \pm x) = \mp \sin x,$$

$$\tan(90° \pm x) = \mp \cot x, \quad \cot(90° \pm x) = \mp \tan x,$$

$$\sin(180° \pm x) = \mp \sin x, \quad \cos(180° \pm x) = -\cos x,$$

$$\tan(180° \pm x) = \pm \tan x, \quad \cot(180° \pm x) = \pm \tan x.$$

Addition and subtraction formulas:

$$\sin(a \pm b) = \sin a \cos b \pm \cos a \sin b,$$

$$\cos(a \pm b) = \cos a \cos b \mp \sin a \sin b,$$

$$\tan(a \pm b) = \frac{\tan a \pm \tan b}{1 \mp \tan a \tan b}.$$

Double-angle formulas:

$$\sin 2a = 2 \sin a \cos a,$$

$$\cos 2a = 2\cos^2 a - 1 = 1 - 2\sin^2 a,$$

$$\tan 2a = \frac{2 \tan a}{1 - \tan^2 a}.$$

Half-angle formulas:

$$\sin a = \frac{2 \tan \frac{a}{2}}{1 + \tan^2 \frac{a}{2}},$$

$$\cos a = \frac{1 - \tan^2 \frac{a}{2}}{1 + \tan^2 \frac{a}{2}},$$

$$\tan a = \frac{2 \tan \frac{a}{2}}{1 - \tan^2 \frac{a}{2}}.$$

Sum-to-product formulas:

$$\sin a + \sin b = 2 \sin \frac{a+b}{2} \cos \frac{a-b}{2},$$

$$\cos a + \cos b = 2 \cos \frac{a+b}{2} \cos \frac{a-b}{2},$$

$$\tan a + \tan b = \frac{\sin(a+b)}{\cos a \cos b}.$$

Difference-to-product formulas:

$$\sin a - \sin b = 2 \sin \frac{a-b}{2} \cos \frac{a+b}{2},$$

$$\cos a - \cos b = -2 \sin \frac{a-b}{2} \sin \frac{a+b}{2},$$

$$\tan a - \tan b = \frac{\sin(a-b)}{\cos a \cos b}.$$

Product-to-sum formulas:

$$2 \sin a \cos b = \sin(a+b) + \sin(a-b),$$

$$2 \cos a \cos b = \cos(a+b) + \cos(a-b),$$

$$2 \sin a \sin b = -\cos(a+b) + \cos(a-b).$$

6
Further Reading

1. Andreescu, T.; Kedlaya, K.; Zeitz, P., *Mathematical Contests 1995-1996: Olympiad Problems from around the World, with Solutions*, American Mathematics Competitions, 1997.

2. Andreescu, T.; Kedlaya, K., *Mathematical Contests 1996–1997: Olympiad Problems from around the World, with Solutions*, American Mathematics Competitions, 1998.

3. Andreescu, T.; Kedlaya, K., *Mathematical Contests 1997–1998: Olympiad Problems from around the World, with Solutions*, American Mathematics Competitions, 1999.

4. Andreescu, T.; Feng, Z., *Mathematical Olympiads: Problems and Solutions from around the World, 1998–1999*, Mathematical Association of America, 2000.

5. Andreescu, T.; Gelca, R., *Mathematical Olympiad Challenges*, Birkhäuser, 2000.

6. Barbeau, E., *Polynomials*, Springer-Verlag, 1989.

7. Beckenbach, E. F.; Bellman, R., *An Introduction to Inequalities*, New Mathematical Library, Vol. 3, Mathematical Association of America, 1961.

8. Chinn, W. G.; Steenrod, N. E., *First Concepts of Topology*, New Mathematical Library, Vol. 27, Mathematical Association of America, 1966.

9. Cofman, J., *What to Solve?*, Oxford Science Publications, 1990.

10. Coxeter, H. S. M.; Greitzer, S. L., *Geometry Revisited*, New Mathematical Library, Vol. 19, Mathematical Association of America, 1967.

11. Doob, M., *The Canadian Mathematical Olympiad 1969–1993*, University of Toronto Press, 1993.
12. Engel, A., *Problem-Solving Strategies*, Problem Books in Mathematics, Springer, 1998.
13. Fomin, D.; Kirichenko, A., *Leningrad Mathematical Olympiads 1987–1991*, MathPro Press, 1994.
14. Fomin, D.; Genkin, S.; Itenberg, I., *Mathematical Circles*, American Mathematical Society, 1996.
15. Graham, R. L.; Knuth, D. E.; Patashnik, O., *Concrete Mathematics*, Addison-Wesley, 1989.
16. Greitzer, S. L., *International Mathematical Olympiads, 1959–1977*, New Mathematical Library, Vol. 27, Mathematical Association of America, 1978.
17. Grossman, I.; Magnus, W., *Groups and Their Graphs*, New Mathematical Library, Vol. 14, Mathematical Association of America, 1964.
18. Kazarinoff, N. D., *Geometric Inequalities*, New Mathematical Library, Vol. 4, Mathematical Association of America, 1961.
19. Klamkin, M., *International Mathematical Olympiads, 1978–1985*, New Mathematical Library, Vol. 31, Mathematical Association of America, 1986.
20. Klamkin, M., *USA Mathematical Olympiads, 1972–1986*, New Mathematical Library, Vol. 33, Mathematical Association of America, 1988.
21. Kürschák, J., *Hungarian Problem Book, volumes I & II*, New Mathematical Library, Vols. 11 & 12, Mathematical Association of America, 1967.
22. Kuczma, M., *144 problems of the Austrian-Polish Mathematics Competition 1978–1993*, The Academic Distribution Center, 1994.
23. Larson, L. C., *Problem-Solving Through Problems*, Springer-Verlag, 1983.
24. Lausch, H. *The Asian Pacific Mathematics Olympiad 1989–1993*, Australian Mathematics Trust, 1994.
25. Lozansky, E.; Rousseau, C. *Winning Solutions*, Springer, 1996.
26. Ore, O., *Graphs and Their Uses*, New Mathematical Library, Vol. 34,

Mathematical Association of America, 1963; revised and updated by Robin Wilson, 1990.

27. Ore, O., *Invitation to Number Theory*, New Mathematical Library, Vol. 20, Mathematical Association of America, 1967.

28. Liu, A., *Chinese Mathematics Competitions and Olympiads 1981–1993*, Australian Mathematics Trust, 1998.

29. Sharygin, I. F., *Problems in Plane Geometry*, Mir, Moscow, 1988.

30. Sharygin, I. F., *Problems in Solid Geometry*, Mir, Moscow, 1986.

31. Shklarsky, D. O; Chentzov, N. N; Yaglom, I. M., *The USSR Olympiad Problem Book*, Freeman, 1962.

32. Slinko, A., *USSR Mathematical Olympiads 1989–1992*, Australian Mathematics Trust, 1997.

33. Soifer, A., *Colorado Mathematical Olympiad: The first ten years*, Center for excellence in mathematics education, 1994.

34. Szekely, G. J., *Contests in Higher Mathematics*, Springer-Verlag, 1996.

35. Stanley, R. P., *Enumerative Combinatorics*, Cambridge University Press, 1997.

36. Taylor, P. J., *Tournament of Towns 1980–1984*, Australian Mathematics Trust, 1993.

37. Taylor, P. J., *Tournament of Towns 1984–1989*, Australian Mathematics Trust, 1992.

38. Taylor, P. J., *Tournament of Towns 1989–1993*, Australian Mathematics Trust, 1994.

39. Taylor, P. J.; Storozhev, A., *Tournament of Towns 1993–1997*, Australian Mathematics Trust, 1998.

40. Tomescu, I., *Problems in Combinatorics and Graph Theory*, Wiley, 1985.

41. Vanden Eynden, C., *Elementary Number Theory*, McGraw-Hill, 1987.

42. Wilf, H. S., *Generatingfunctionology*, Academic Press, 1994.

43. Wilson, R., *Introduction to Graph Theory*, Academic Press, 1972.

44. Yaglom, I. M., *Geometric Transformations*, New Mathematical Library, Vol. 8, Mathematical Association of America, 1962.

45. Yaglom , I. M., *Geometric Transformations II*, New Mathematical Library, Vol. 21, Mathematical Association of America, 1968.

46. Yaglom , I. M., *Geometric Transformations III*, New Mathematical Library, Vol. 24, Mathematical Association of America, 1973.

47. Zeitz, P., *The Art and Craft of Problem Solving*, John Wiley & Sons, 1999.

Appendix

1 2000 Olympiad Results

The top twelve students on the 2000 USAMO were (in alphabetical order):

David G. Arthur	Toronto, ON
Reid W. Barton	Arlington, MA
Gabriel D. Carroll	Oakland, CA
Kamaldeep S. Gandhi	New York, NY
Ian Le	Princeton Junction, NJ
George Lee, Jr.	San Mateo, CA
Ricky I. Liu	Newton, MA
Po-Ru Loh	Madison, WI
Po-Shen Loh	Madison, WI
Oaz Nir	Saratoga, CA
Paul A. Valiant	Belmont, MA
Yian Zhang	Madison, WI

Reid Barton and Ricky Liu were the winners of the Samuel Greitzer-Murray Klamkin award, given to the top scorer(s) on the USAMO. The Clay Mathematics Institute (CMI) award, to be presented for a solution of oustanding elegance, and carrying a $1000 cash prize, was presented to Ricky Liu for his solution to USAMO Problem 3.

The USA team members were chosen based on their combined performance on the 29th annual USAMO and the Team Selection Test that took place at this year's MOSP held at the University of Nebraska-Lincoln, June 6- July 4, 2000. Members of the USA team at the 2000 IMO (Taejon, Republic of Korea) were Reid Barton, George Lee, Ricky Liu, Po-Ru Loh, Oaz Nir, and Paul Valiant. Titu Andreescu (Director of the

American Mathematics Competitions) and Zuming Feng (Phillips Exeter Academy) served as team leader and deputy leader, respectively. The team was also accompanied by Dick Gibbs (Chair, Committee on the American Mathematics Competitions, Fort Lewis College), as the official observer of the team leader.

At the 2000 IMO, gold medals were awarded to students scoring between 30 and 42 points (there were 4 perfect papers on this very difficult exam), silver medals to students scoring between 20 and 29 points, and bronze medals to students scoring between 11 and 19 points. Barton's 39 tied for 5th. The team's individual performances were as follows:

Barton	Homeschooled	GOLD Medalist
Lee	Aragon HS	GOLD Medalist
Liu	Newton South HS	SILVER Medalist
P.-R. Loh	James Madison Memorial HS	SILVER Medalist
Nir	Monta Vista HS	GOLD Medalist
Valiant	Milton Academy	SILVER Medalist

In terms of total score (out of a maximum of 252), the highest ranking of the 82 participating teams were as follows:

China	218	Belarus	165
Russia	215	Taiwan	164
USA	184	Hungary	156
Korea	172	Iran	155
Bulgaria	169	Israel	139
Vietnam	169	Romania	139

The 2001 IMO is scheduled to be held in Washington, DC and Fairfax, VA. Contact Walter Mientka at walter@amc.unl.edu or Kiran Kedlaya at kedlaya@math.berkeley.edu for more information about the 2001 IMO.

The 2000 USAMO was prepared by Titu Andreescu (Chair), Zuming Feng, Kiran Kedlaya, Alexander Soifer, Richard Stong and Zvezdelina Stankova-Frenkel. The Team Selection Test was prepared by Titu Andreescu and Kiran Kedlaya. The MOSP was held at the University of Nebraska, Lincoln. Titu Andreescu (Director), Zuming Feng, Razvan Gelca, Kiran Kedlaya, and Zvezdelina Stankova-Frenkel served as instructors, assisted by Melanie Wood and Daniel Stronger.

For more information about the USAMO or the MOSP, contact Titu Andreescu at titu@amc.unl.edu.

2 1999 Olympiad Results

The top eight students on the 1999 USAMO were (in alphabetical order):

Reid W. Barton	Arlington, MA
Gabriel D. Carroll	Oakland, CA
Lawrence O. Detlor	New York, NY
Stephen E. Haas	Sunnyvale, CA
Po-Shen Loh	Madison, WI
Alexander B. Schwartz	Bryn Mawr, PA
Paul A. Valiant	Belmont, MA
Melanie E. Wood	Indianapolis, IN

Alexander (Sasha) Schwartz was the winner of the Samuel Greitzer-Murray Klamkin award, given to the top scorer(s) on the USAMO. Newly introduced this year was the Clay Mathematics Institute (CMI) award, to be presented (at the discretion of the USAMO graders) for a solution of oustanding elegance, and carrying a $1000 cash prize. The CMI award was presented to Po-Ru Loh (Madison, WI; brother of Po-Shen Loh) for his solution to USAMO Problem 2.

Members of the USA team at the 1999 IMO (Bucharest, Romania) were Reid Barton, Gabriel Carroll, Lawrence Detlor, Po-Shen Loh, Paul Valiant, and Melanie Wood. Titu Andreescu (Director of the American Mathematics Competitions) and Kiran Kedlaya (Massachusetts Institute of Technology) served as team leader and deputy leader, respectively. The team was also accompanied by Walter Mientka (University of Nebraska, Lincoln), who served as secretary to the IMO Advisory Board and as the official observer of the team leader.

At the 1999 IMO, gold medals were awarded to students scoring between 28 and 39 points (39 was the highest score obtained on this extremely difficult exam), silver medals to students scoring between 19 and 27 points, and bronze medals to students scoring between 12 and 18 points. Barton's 34 tied for 13th. The team's individual performances were as follows:

Barton	GOLD Medalist
Carroll	SILVER Medalist
Detlor	BRONZE Medalist
P.-S. Loh	SILVER Medalist
Valiant	GOLD Medalist
Wood	SILVER Medalist

In terms of total score, the highest ranking of the 81 participating teams were as follows:

China	182	Korea	164
Russia	182	Iran	159
Vietnam	177	Taiwan	153
Romania	173	USA	150
Bulgaria	170	Hungary	147
Belarus	167	Ukraine	136

The 1999 USAMO was prepared by Titu Andreescu (Chair), Zuming Feng, Kiran Kedlaya, Alexander Soifer and Zvezdelina Stankova-Frenkel. The MOSP was held at the University of Nebraska, Lincoln. Titu Andreescu (Director), Zuming Feng, Kiran Kedlaya, and Zvezdelina Stankova-Frenkel served as instructors, assisted by Andrei Gnepp and Daniel Stronger.

3 1998 Olympiad Results

The top eight students on the 1998 USAMO were (in alphabetical order):

Reid W. Barton	Arlington, MA
Gabriel D. Carroll	Oakland, CA
Kevin D. Lacker	Cinncinati, OH
Alexander B. Schwartz	Bryn Mawr, PA
David Speyer	Wallingford, CT
Paul A. Valiant	Belmont, MA
David Vickrey	Vermillion, SD
Melanie E. Wood	Indianapolis, IN

Alexander (Sasha) Schwartz and Melanie Wood tied as winners of the Samuel Greitzer-Murray Klamkin award, given to the top scorer(s) on the USAMO.

Members of the USA team at the 1998 IMO (Taipei, Taiwan) were Reid Barton, Gabriel Carroll, Kevin Lacker, Alexander Schwartz, Paul Valiant, and Melanie Wood. Titu Andreescu (Illinois Mathematics and Science Academy) and Elgin Johnston (Iowa State University) served as team leader and deputy leader, respectively. The team was also accompanied by Walter Mientka (University of Nebraska, Lincoln), who served as secretary to the IMO Advisory Board and as the official observer of the team leader.

At the 1998 IMO, gold medals were awarded to students scoring between 31 and 42 points, silver medals to students scoring between 24 and 30 points, and bronze medals to students scoring between 14 and 23 points. Schwartz's 36 tied for 12th. The team's individual performances were as follows:

Barton	GOLD Medalist
Carroll	GOLD Medalist
Lacker	SILVER Medalist
Schwartz	GOLD Medalist
Valiant	SILVER Medalist
Wood	SILVER Medalist

In terms of total score, the highest ranking of the 76 participating teams were as follows:

Iran	211		Russia	175
Bulgaria	195		India	174
USA	186		Ukraine	166
Hungary	186		Vietnam	158
Taiwan	184		Yugoslavia	156

The 1998 USAMO was prepared by Titu Andreescu (Chair), Elgin Johnston, Jim Propp, Alexander Soifer, Richard Stong and Paul Zietz. The MOSP was held at the University of Nebraska, Lincoln. Titu Andreescu (Director), Zuming Feng, Razvan Gelca, Elgin Johnston, Kiran Kedlaya, and Zvezdelina Stankova-Frenkel served as instructors, assisted by Carl Bosley and Noam Shazeer.

4 1997 Olympiad Results

The top eight students on the 1997 USAMO were (in alphabetical order):

Carl J. Bosley	Topeka, KS
Li-Chung Chen	Cupertino, CA
John J. Clyde	New Plymouth, ID
Nathan G. Curtis	Alexandria, VA
Kevin Lacker	Cinncinati, OH
Devesh Maulik	Roslyn Heights, NY
Josh P. Nichols-Barrer	Newton Center, MA
Daniel P. Stronger	New York, NY

Josh P. Nichols-Barrer was the winner of the Samuel Greitzer-Murray Klamkin award, given to the top scorer(s) on the USAMO.

Members of the USA team at the 1997 IMO (Mar del Plata, Argentina) were Carl Bosley, Li-Chung Chen, John Clyde, Nathan Curtis, Josh Nichols-Barrer and Daniel Stronger. Titu Andreescu (Illinois Mathematics and Science Academy) and Elgin Johnston (Iowa State University) served as team leader and deputy leader, respectively. The team was also accompanied by Walter Mientka (University of Nebraska, Lincoln), who served as secretary to the IMO Advisory Board and as the official observer of the team leader.

At the 1997 IMO, gold medals were awarded to students scoring between 35 and 42 points, silver medals to students scoring between 25 and 34 points, and bronze medals to students scoring between 15 and 24 points. Bosley scored one of the four perfect papers in the contest. Curtis's 40 tied for 5th. The team's individual performances were as follows:

Bosley	GOLD Medalist
Chen	SILVER Medalist
Clyde	SILVER Medalist
Curtis	GOLD Medalist
Nichols-Barrer	SILVER Medalist
Stronger	SILVER Medalist

In terms of total score, the highest ranking of the 82 participating teams were as follows:

China	223	Ukraine	195
Hungary	219	Bulgaria	191
Iran	217	Romania	191
USA	204	Australia	187
Russia	204	Vietnam	183

The 1997 USAMO was prepared by Titu Andreescu, Elgin Johnston, Jim Propp, Cecil Rousseau (Chair), Alexander Soifer, Richard Stong and Paul Zietz. The MOSP was held at the University of Nebraska, Lincoln. Titu Andreescu (Director), Fan Chung, Zuming Feng, Razvan Gelca, Elgin Johnston, and Kiran Kedlaya served as instructors, assisted by Jeremy Bem and Jonathan Weinstein.

5 1996 Olympiad Results

The top eight students on the 1996 USAMO were (in alphabetical order):

Appendix

Carl J. Bosley	Topeka, KS
Christopher Chang	Palo Alto, CA
Nathan G. Curtis	Alexandria, VA
Michael R. Korn	Arden Hills, MN
Carl A. Miller	Bethesda, MD
Josh P. Nichols-Barrer	Newton Center, MA
Alexander H. Saltman	Austin, TX
Daniel P. Stronger	New York, NY

Christopher Chang was the winner of the Samuel Greitzer-Murray Klamkin award, given to the top scorer(s) on the USAMO.

Members of the USA team at the 1996 IMO (Mumbai, India) were Carl Bosley, Christopher Chang, Nathan Curtis, Michael Korn, Carl Miller, and Alexander Saltman. Titu Andreescu (Illinois Mathematics and Science Academy) and Kiran Kedlaya (Princeton University) served as team leader and deputy leader, respectively. The team was also accompanied by Walter Mientka (University of Nebraska, Lincoln), as the official observer of the team leader.

At the 1996 IMO, gold medals were awarded to students scoring between 28 and 42 points, silver medals to students scoring between 20 and 27 points, and bronze medals to students scoring between 12 and 19 points. There was one perfect paper on this extremely difficult exam (by Ciprian Manolescu, Romania. Manolescu participated in IMO in 1995, 1996, 1997, and he submitted a perfect paper in each contest). Saltman's 37 and Chang's 36 tied for 4th and 7th, respectively. The team's individual performances were as follows:

Bosley	GOLD Medalist
Chang	GOLD Medalist
Curtis	SILVER Medalist
Korn	GOLD Medalist
Miller	SILVER Medalist
Saltman	GOLD Medalist

In terms of total score, the highest ranking of the 75 participating teams were as follows:

Romania	187	Vietnam	155
USA	185	Korea	151
Hungary	167	Iran	143
Russia	162	Germany	137
United Kingdom	161	Bulgria	136
China	160	Japan	136

The 1996 USAMO was prepared by Titu Andreescu, Elgin Johnston, Jim Propp, Cecil Rousseau (Chair), Alexander Soifer, Richard Stong and Paul Zietz. The MOSP was held at the University of Nebraska, Lincoln. Titu Andreescu (Director), Elgin Johnston, Kiran Kedlaya, and Paul Zeitz served as instructors, assisted by Jeremy Bem and Jonathan Weinstein.

6 1996–2000 Cumulative IMO Results

In terms of total scores (out of a maximum of 1260 points), the highest ranking of the participating IMO teams is as follows:

Russia	938	Romania	845
USA	909	Vietnam	842
Iran	885	Korea	805
Hungary	875	China[1]	783
Bulgaria	861	Taiwan	749

In all, more and more countries now value the crucial role of meaningful problem solving in mathematics education. The competition is getting tougher and tougher. A top ten finish is not a given for traditional powerhouses anymore.

---

[1] China did not participate in the 1998 IMO (Taiwan)

Titu Andreescu is the Director of the American Mathematics Competitions, a program of the Mathematical Association of America (MAA). He also serves as Chair of the USA Mathematical Olympiad Committee, Head Coach of the USA International Mathematical Olympiad Team, and Director of the Mathematical Olympiad Summer Program.

Before joining the staff of the MAA, Titu was an instructor of mathematics at the Illinois Mathematics and Science Academy (IMSA) in Aurora, Illinois (1991–1998). From 1981 through 1989, Titu served as Professor of Mathematics at Loga Academy in Timisoara, Romania. While living in Romania, he was appointed Counselor to the Romanian Ministry of Education and served as Editor-in-Chief of Timisoara's *Mathematical Review*.

Titu received the Distinguished Teacher Award from the Romanian Ministry of Education in 1983 and the Edyth May Sliffe Award for Distinguished High School Mathematics Teaching from the MAA in 1994.

Zuming Feng graduated with a PhD degree from Johns Hopkins University with an emphasis on Algebraic Number Theory and Elliptic Curves. He teaches at Phillips Exeter Academy. Zuming is a coach of the USA International Mathematical Olympiad (IMO) Team, a member of the USA Mathematical Olympiad Committee, and an assistant director of the USA Mathematical Olympiad Summer Program. He received the Edyth May Sliffe Award for Distinguished High School Mathematics Teaching from the MAA in 1996.